Conservation of
Fuel and Power

2006 Edition

Building BPG
Performance
Group

bre

© NBS, 2006
Published by NBS, The Old Post Office, St Nicholas Street, Newcastle Upon Tyne
NE1 1RE

ISBN-10 1 85946 199 9
ISBN-13 978 1 85946 199 0

Stock Code 57270

British Library Cataloguing in Publications Data
A catalogue record for this book is available from the British Library.

Publisher: Steven Cross
Commissioning Editor: Matthew Thompson
Project Editor: Anna Walters
Editor: Prepress Projects
Designed by Philip Handley
Typeset by Prepress Projects
Printed and bound by Hobbs the Printer, Hampshire

NBS is a part of RIBA Enterprises Ltd. www.ribaenterprises.com

Contents

Foreword

Designing energy-efficient buildings has been a growing preoccupation for many architects, prompted initially by the oil crisis of the 1970s and latterly by the growing understanding of the link between increasing CO_2 emissions from burning fossil fuels and global warming. Given the significance of CO_2 emissions from buildings as a proportion of our national total, it is unsurprising that the government has identified the sector as a key element in its overall reduction strategy. The new four-part version of Part L is perhaps the most obvious evidence of its ongoing commitment to increasingly stringent legislation in this area to compel the industry to raise its standards of thermal performance and to promote the use of low and zero carbon technologies.

The new regulations are very different from those that the industry has been accustomed to. They expand the scope of work that comes under their control, particularly when working with existing buildings. They call for more testing to demonstrate that design intentions have been carried through into the final build, and certification that services have been properly commissioned. They also, and perhaps more importantly, adopt a fundamentally different approach to demonstrating compliance, both at the design stage and on completion of the building, based on a calculation, largely computer software based, of the CO_2 emissions of the building as being the sole acceptable method. This eliminates the more user-friendly options with which the industry is familiar and, perhaps as a consequence, the Approved Documents have taken a step back from a 'how to' manual, which they became when the illustrated A4 format was introduced, into a somewhat opaque document setting out the procedure for demonstrating compliance.

Thankfully there is now a guide that unpicks these unfamiliar and rather tortuous new requirements. Designed to serve the needs of the designer, the **Guide to Part L** comprises a series of invaluable flowcharts for each of the four new documents, all of which are backed up with guidance that amplifies important points and clarifies meanings. This is supported by extensive appendices that explain some of the science, concepts and processes that are required in order to comply. Usefully, there is also a comprehensive bibliography of not only the second- and third-tier documents referred to in Part L but also many others that are central to good practice. I can see it becoming a genuinely important reference work for architects, construction professionals and even building control officers as they grapple with the new approach. I thoroughly recommend it to you.

Bill Gething
of Feilden Clegg Bradley LLP, Chair of the RIBA Sustainable Futures Committee and the RIBA President's Advisor on Sustainability
June 2006

General introduction to L1A, L1B, L2A and L2B

1.0 Introduction to the guide

The busy designer, whether a sole practitioner or a team member in a large global organisation, may not have time to study and understand in detail all the relevant legislation. This guide is intended to be at the designer's side to provide an immediate source of advice when time is precious and deadlines have to be met.

The layout provides quick access to the essentials so that they can be assimilated easily without cluttering the text with superfluous detail. The following two types of highlighting are used.

 ESSENTIAL

> To aid rapid digestion of the key points, boxes like this contain mandatory requirements

Readers requiring more details and explanation of a particular area of interest can find these in the appendices.

 FOR INFORMATION

> Background information and points of interest that can be skipped over are in boxes like this

Words highlighted in **bold** have a specific technical meaning that is explained in the Glossary section of this guide.

The requirements of the Part L Approved Documents (ADs) are difficult to assimilate and apply. The authors have spent many man-hours – on the reader's behalf – looking into the meaning and importance of each clause in the ADs, discussing these with the publishers, researching the cross-references (some of which are as yet unpublished) and compiling the guidance. Where cross-reference documents are not available, the guide aims to either fill in the gaps or provide alternative solutions.

There is quite a lot of repetition in the four sections that constitute the Part L ADs, so this guide is designed to give a quick 'road map' overview of each section. The detailed guidance on any particular subject is mostly to be found in the appendices.

There are some caveats. Some of the cross-reference documents are still in draft and unpublished; where this is the case, we have highlighted this. The guide cannot cover every conceivable design approach, so where we have included 'deemed to satisfy' solutions, they are fairly traditional. Highly specialised or individual approaches to meeting the requirements of the Regulations are permitted subject to satisfying the criteria, but we have not attempted to cover them here.

Every effort has been made to ensure the accuracy of the information included in this guide at the time of publication; however, legislation is constantly being revised and updated, and there are likely to be amendments to the ADs themselves. Readers will have ultimate responsibility for ensuring that they have taken account of all the most up to date requirements of the appropriate legislation.

1.1   Background

The revision of the Part L Approved Documents is a direct outcome of the Government's Energy White Paper[1] commitment to raise the energy performance of buildings. It requires implementation of the European Energy Performance of Buildings Directive to Member States setting performance standards requiring certification of buildings, including periodic inspections and certification of air-conditioning and boiler systems.

1.2   What the Building Regulations actually require

Generally, the 'Building Regulations are made for specific purposes: health and safety, energy conservation and the welfare and convenience of disabled people'.[2]

The Regulations that came into force in April 2006 require that reasonable provision should be made for the conservation of fuel and power in buildings which use energy to condition the indoor climate[3] by:

- limiting heat gains and losses through the fabric of the building;
- limiting heat losses from pipes, ducts and vessels used for space heating, space cooling and hot water storage;
- providing energy-efficient fixed building services which have effective controls and which have been properly commissioned;
- providing the owner with sufficient information about the building and building services and their maintenance so that it can be operated to minimise use of fuel and power as much as is reasonable in the circumstances.

Notes

1. *Our Energy Future: Creating A Low Carbon Economy.* Presented to Parliament by the Secretary of State for Trade and Industry by command of Her Majesty, February 2003.

2. Approved Documents: Use of Guidance, Technical Specifications (reproduced in each Part).

3. Refer to 'The Requirement' in each of the four sections of Part L.

The Approved Documents indicate clearly what these limitations and requirements are as well as what the reasonable provisions are and how they can be met.

Note that the requirements of other parts of the Regulations must still be satisfied when meeting the requirements of Part L (2006). Part F (2006), for instance, describes requirements for ventilation which should not conflict with the requirements for limiting air leakage under Part L. In certain circumstances, alterations to the structure of a building under Part A, fire protection under Part B or access to and use of buildings under Part M are **material alterations** (see L1B paragraph 25; L2B paragraph 37), which will mean that improvements to thermal performance will be required under Part L. Where **controlled services or fittings** (see L1B paragraph 27; L2B paragraph 39) are specified, the relevant requirements of Parts G, H, J and P also have to be considered.

1.3 Changes since previous edition

The changes are set out at the beginning of each of the four sections of Part L under the headings 'Other changes to the Regulations' and 'New Part VA to the Regulations'.[4]

The new Regulation L1 (2006) requires all buildings to be energy efficient and to have their building services fitted with effective controls and be properly commissioned. In addition, the householder or building owner needs to be given sufficient information on the building, the fixed building services and their maintenance requirements so that the building can be operated efficiently, using no more fuel and power 'than is reasonable in the circumstances'.

The 2000 version of Regulation L1 focused on limiting heat loss by defining levels of insulation; it neither referred to heat gains nor required that the building services were energy efficient and properly commissioned. There was no requirement to provide information to users, so that the building could be operated efficiently.

In April 2002, Regulation L1 was revised and Regulation L2 was introduced for buildings other than dwellings. These revisions introduced the need for building services to be energy efficient and, in the case of L2, required heat gains to both the building and cooling plant installation to be limited. L2 also required the provision of information on the relevant services installation to enable its operation and maintenance so as to use no more energy than is reasonable.

The April 2006 requirement under Part L is now the same for all building types. The AD though is in four sections to differentiate between dwellings and other buildings, and to further differentiate between new building work and work to existing buildings of each group.

Notes

4. Refer also to Annex D of ODPM Circular 03/2006 'The Building Act 1984', 15 March 2006.

The performance of a building now has to be calculated by reference to the mass of carbon dioxide emitted per square metre over its life (Regulation 17). In the case of dwellings, the intention is that the predicted performance will form part of the **Home Information Pack**.

For all new buildings, the calculated 'dwelling emission rate' for dwellings and the 'building emission rate' for other types of building based on the actual building design are no greater than the calculated 'target emission rate'. There is also reliance on 'second-tier' and 'third-tier' documents, which provide the specific requirements and good practice guidance respectively.

Compliance checklists are provided to show who should produce evidence of compliance to the relevant Building Control body. There is greater emphasis placed on 'approved competent persons' to self-certify or produce evidence of compliance with the Regulations, and the intention is that eventually 'accredited details' can be used by a builder to demonstrate compliance and reduce the number of post-construction tests required.

If there is any change in the energy status of a building, the Regulations now require that reasonable provision should be made to improve the energy performance of thermal elements that are being renovated or renewed (Regulation 4A), even though they may be neither part of a material alteration nor regarded as 'building work' as defined in Regulation 3. As such, they do not need to comply with the other technical requirements of the Regulations (Parts A–P inclusive). Refer also to Schedule 2A of the Regulations reproduced near the front of each of the four sections of Part L. Regulation 17D requires **consequential improvement** to existing buildings where large extensions (> 1000m²) are proposed.

Regulation 20B makes it mandatory that new buildings, including dwellings, are tested for air leakage to demonstrate that the construction meets the design assumptions. This greatly extends the requirements under the previous Regulation 18, which made testing an option.

Regulation 20C requires a formal certificate to show that commissioning of heating and hot water systems has been carried out by an approved Part J or Part P competent person in the case of dwellings or by a suitably qualified person in the case of other building types. It should be noted that the checklists need to be completed at both design and post-construction stages.

Since Part L2 (2002), the units have changed from carbon to carbon dioxide emission (to be more in line with the directive). This means that the factors used for CO_2 emissions appear different. (They have changed by a ratio of 44/12![5])

Notes   5. Paragraph 23 of Part L2A.

1.4   Which section of Part L?

The following is a brief summary to help readers to find the appropriate section of the guide quickly. It may also help to refer to Table 1 of this guide. It should be noted that conservatories under 30m² are outside the scope of the Regulations generally and are exempt from these requirements. There are special considerations in the form of relaxations for **historic buildings** or buildings in conservation areas, national parks or areas of outstanding natural beauty.[6] There are also self-certification schemes for notifiable work, and certain small-scale works are non-notifiable. These are listed in Schedule 2A and Schedule 2B reproduced in the early part of each of the four sections of Part L.

> If you are modifying but not extending an existing building, then *any* new or replacement window, rooflight, roof window or external door with more than 50% of internal area glazed and *any* new space heating, hot water service boiler, air-conditioning or mechanical ventilation in an existing building is within the remit of these Regulations. It also includes the requirement in certain circumstances to improve the energy performance when replacing or renovating walls, floors or roofs.

Approved Document L1A Conservation of Fuel and Power in New Dwellings (2006 edition) applies to new-build **dwellings** such as houses and flats.

Approved Document L1B Conservation of Fuel and Power in Existing Dwellings covers **extensions** and **material alterations** to existing dwellings, as well as **dwellings** created by a **material change of use**. It also applies to work on certain services or fittings (called **controlled services or fittings**) such as new external doors or windows, new or replacement hot water systems, mechanical ventilation and cooling systems, insulation of existing pipes and ducts, and rewired or extended lighting circuits as well as the provision of a new, renovation, or upgrade of a **thermal element**.

Approved Document L2A Conservation of Fuel and Power in New Buildings other than Dwellings covers new buildings other than dwellings, including the first **fit-out works** to a building.

Approved Document L2B Conservation of Fuel and Power in Existing Buildings other than Dwellings covers **consequential improvements** (required because you are adding an extension or changing a service to a building with a useful area of more than 1000m²), **extensions, material alterations** and **material change of use** to existing buildings, replacement of certain services or fittings (called **controlled services or fittings**) such

---

Notes

6.   *Building Regulations and Historic Buildings*, English Heritage, September 2002.

as new external doors or windows, new hot water systems, mechanical ventilation and cooling systems, insulation of existing pipes and ducts, and lighting, as well as the provision of a new renovation or upgrade of a thermal element (wall, roof or floor).

> Take care because L2A also covers heated common areas in residential buildings as well as nursing homes and student accommodation and – in mixed use development – the commercial or retail space. It also includes extensions over 100m² AND greater than 25% of the existing floor area.

Table 1 lists the appropriate Part L for the range of options that are described within the approved document.

If you follow the guidance in the Approved Documents Part L, this would fulfil the requirements of the Building Regulations in the following clauses:

- 4A (renovated or replaced thermal elements to comply with requirements);
- 4B (where there is a change of energy status);
- 17A (the requirement for a methodology to calculate the energy performance of a building);
- 17B (the requirement to set minimum energy performance requirements in the form of target CO_2 emission);
- 17C (any new building shall meet the target CO_2 emission rate);
- 17D (extensions or provision of new, or increased capacity to, fixed building services in buildings over 1000m² are to comply with the requirements of Part L if technically and economically feasible);
- 20B (ensure, give notice and confirm that pressure testing has been carried out and recorded);
- 20C (provide notice by a certain date that all fixed building services have been properly commissioned);
- and 20D (provide a notice to the local authority which specifies the target CO_2 emission rate for the building and the calculated CO_2 rate for the building as constructed).

Table 1
Which part of Part L?

New build	Existing	Use of space	Use Part
Dwelling		House or flat	L1A
	Extension to dwelling	House or flat	L1B
Commercial space or office in a dwelling that could revert to domestic use		If there is direct access from the dwelling to the commercial space and both are within the same thermal envelope and the living space is a substantial proportion of the whole area of the building	L1A
New building other than dwellings		Not a house or flat. Refer below in this table for buildings that are exempt	L2A
	New non-dwelling extension of more than 100m² AND more than 25% existing gross floor area	Not a house, a flat or a room used as a dwelling. Refer below in this table for buildings that are exempt	L2A
	New non-dwelling extension less than 100m²	Not a house, a flat or a room used as a dwelling. Refer below in this table for buildings that are exempt	L2B
	New non-dwelling extension less than 25% existing gross floor area	Not a house, a flat or a room used as a dwelling. Refer below in this table for buildings that are exempt	L2B
	Dwelling created as part of a material change of use	House or flat	L1B
	Provision of 'controlled fittings'	House or flat	L1B
	Provision of 'controlled services'	House or flat	L1B
	Material alteration to existing dwelling	House or flat	L1B
	Renovation of thermal element	House or flat	L1B

Table 1
Continued

New build	Existing	Use of space	Use Part
Unheated common parts		Flats only	L1A, fabric to paragraphs 33–36
	Unheated common parts	Flats only	L1B
Unheated common parts		Where common parts serve a mixed development that includes flats	L2A
	Unheated common parts	Where common parts serve a mixed development that includes flats	L2B
Heated common parts		Flats only	L2A
Heated common parts		Where common parts serve a mixed development which includes flats	L2A
	Heated common parts	Where common parts serve a mixed development, including flats; all building types including flats	L2B
Nursing home		Accommodation	L2A
	Nursing home	Accommodation	L2B
Student accommodation		Accommodation	L2A
	Student accommodation	Accommodation	L2B
Hostel		Accommodation	L2A
	Hostel	Accommodation	L2B
Hotel		Accommodation	L2A
	Hotel	Accommodation	L2B
*Conservatory, area less than 30m²	*Conservatory, area less than 30m²	Any	Exempt
*Conservatory, area greater than 30m²	*Conservatory, area greater than 30m²	Attached to, but outside, thermal envelope of house or flat	L1B

New build	Existing	Use of space	Use Part
*Conservatory, area greater than 30 m²	*Conservatory, area greater than 30m²	Attached to, but outside, thermal envelope of any building other than dwelling	L2B
	Enclosing an existing courtyard or under an extended roof	Not a **dwelling**	L2B
	Historic monument or building where compliance with Part L would unacceptably alter its character or appearance	Refer to guidance notes in L1A and L2A for specific definitions of historic monuments or buildings that are exempt	Exempt
Places of worship such as churches, mosques, synagogues and temples	Places of worship such as churches, mosques, synagogues and temples	This does not include 'buildings used for religious activity' that have additional functions	Exempt
Temporary building with a planned life of less than 2 years			Exempt but follow *Energy Performance Standards for Modular Buildings*
Modular building constructed of more than 70% prefabricated sub-assemblies from centrally held or disassembled stock manufactured before April 2006			Exempt but follow *Energy Performance Standards for Modular Buildings*
Modular buildings		Other than those described in the preceding two rows	L2A
Industrial sites with a low energy demand	Industrial sites with a low energy demand	The regulations do not cover the energy consumed directly by a commercial or industrial process	Exempt
Non-residential agricultural buildings with a low energy demand	Non-residential agricultural buildings with a low energy demand		Exempt

Table 1
Continued

New build	Existing	Use of space	Use Part
Residential buildings used less than 4 months of the year			L1A
	Residential buildings used less than 4 months of the year		L1B
Stand-alone buildings with an area of less than 50m²	Stand-alone buildings with an area of less than 50m²	Not a **dwelling**	Exempt
Stand-alone **dwelling** with an area of less than 50m²			L1A
	Stand-alone **dwelling** with an area of less than 50m²		L2B
Initial **fit-out works**		When completion certificate of building other than dwelling was based on guidance of L2A (2006)	L2A

*A conservatory is an extension that has more than 75% roof area and more than 50% of its external wall made of translucent material. The wall between the conservatory and the dwelling must have a thermal performance equal to that provided for the rest of the dwelling. If one or other of these is not the case, then it may be treated as a 'substantially glazed extension'; however, in most cases, it is not a conservatory and is to be treated as part of the building.

Part L1A: new dwellings

Approved Document L1A Conservation of Fuel and Power in New Dwellings (2006 edition) applies to new-build **dwellings** such as houses and flats (or if attached to a dwelling a space for commercial use that could at some stage revert to being a dwelling; see paragraph 4 in L2A).

2.1   The five criteria

All five criteria have to be met. Energy-efficient measures should not compromise the other requirements of the Building Regulations.

The criteria are:

1. predicted rate of carbon dioxide emissions (**dwelling emission rate**) should be less than the **target emission rate**;
2. thermal performance of building fabric and performance of fixed building services should be no worse than the design limits set out in Part L;
3. passive solar control measures should be provided;
4. as-built performance should be consistent with the predictions;
5. easily understood guidance to occupiers on energy-efficient operation of their systems should be provided.

2.2 A simple road map for compliance

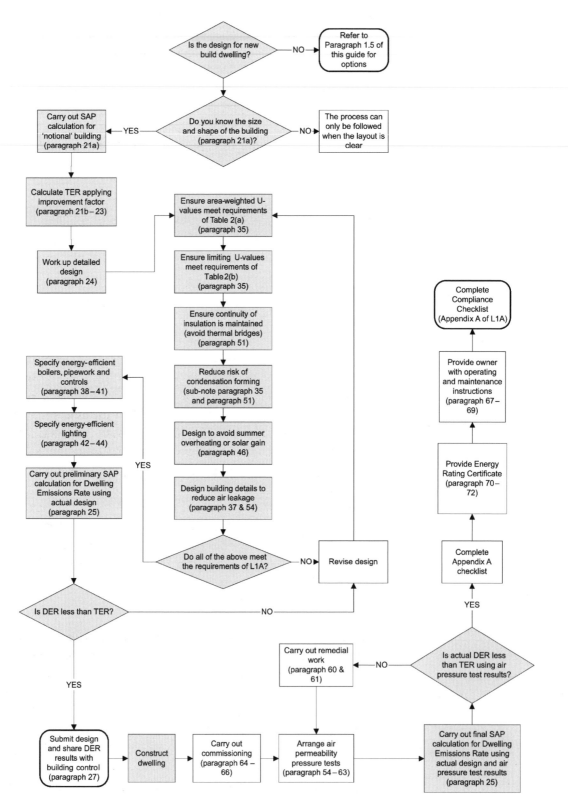

Figure 1
Road map for meeting the provisions of Part L1A

By following this map, you will gather enough information to complete the checklist that is included in the Approved Document L1A as Appendix A. Paragraph A6 in Appendix A states that the CO_2 target is mandatory, but the other checks in the list represent **'reasonable provision'**.

The checklist includes items that are to be produced by 'the builder'; however, in practice, the designer would have to design and specify solutions which meet the requirements in order for the builder to meet the checklist requirements.

The process outlined below follows the order that the designer is likely to follow, rather than the order of the criteria in L1A.

2.2.1 Work out your target emissions (target carbon dioxide emission rate, TER)

Criterion 1: Carbon emissions

TER is expressed as the mass of carbon dioxide in kilograms emitted per square metre of floor area per year as the result of the provision of heating, cooling, hot water, ventilation and internal fixed lighting assuming a standardised household (i.e. using the proposed area and shape of the intended dwelling but using default standard values for the energy performance of fabric and services). Target emissions are the minimum energy performance required to meet the Regulations.

Working out the TER is a two-stage calculation. The first stage is to calculate the CO_2 emissions from heating, cooling, ventilation and lighting in a notional dwelling of the same size and shape as the proposed **dwelling.** The second stage is to calculate the TER by applying the formula in L1A paragraph 21(b). (Details of the information required and how this is actually done is included in Appendix 6.1.)

If an individual dwelling is smaller than 450m² (as most dwellings are), the Government's Standard Assessment Procedure (SAP) must be used for the first part of this calculation. The text, tables and worksheet can be downloaded from http://projects.bre.co.uk/sap2005. It is possible for the calculations to be carried out by hand using the downloaded printed tables, but the use of Government-approved SAP software is likely to be preferred by the designer.

If, on the other hand, each individual dwelling is greater than 450m², then the Simplified Building Energy Model must be used for the first part of the calculation. This is a free download available from http://www.ncm.bre.co.uk/download.jsp. It is not possible to carry out the calculations without the software.

> Note that, on completion of the building, the builder will be required to submit to building control a SAP rating[7]. The information for this purpose is provided by an **authorised SAP assessor**, and it may be appropriate to employ this expert for all the calculations required under Part L.

Please refer to Appendix 6.1 for details.

2.2.2   Ensure U-values comply

Criterion 2: Thermal performance of fabric and services

Table 2 in L1A gives U-values for the area-weighted average (column (a)) as well as for the individual element (column (b)). It is important to check that all U-values in the design meet these minimum levels; however, it is likely that, in order to meet the TER, these U-values will have to be bettered.

It is also important to check details and calculations to rule out the risk of interstitial condensation (BR 262).[8]

Please refer to Appendix 6.6 for details.

2.2.3   Ensure that thermal bridges comply

Criterion 2: Thermal performance of fabric and services

At the internal surfaces, where insulation is discontinuous or is insufficient (window frames and reveals, at the eaves and so on), the effectiveness of the insulation provided is reduced and cool spots are created, where the local inside surface temperature falls below the dewpoint, causing condensation to form on the inner surface. This brings with it a host of problems, including wood rot, crumbling plaster, mildew formation and so on. Careful attention to detailing is required to reduce the risk of thermal bridging.

Guidance on detailing can be found in *Limiting Thermal Bridging and Air Leakage: Robust Construction Details for Dwellings and Similar Buildings*.[9]

Notes

7. Work to design the Energy Performance Certificate required by the EU Directive has been undertaken by FAERO on behalf of BRE (the Building Research Establishment), which maintains the SAP for DEFRA. Refer to www.faero.org.uk for latest design proposals.

8. BRE Report 262 *Thermal Insulation: Avoiding Risks*, 2002 edn.

9. *Limiting Thermal Bridging and Air Leakage: Robust Construction Details for Dwellings and Similar Buildings*, Amendment 1, TSO, 2002.

BR 262[10] shows how thermal bridging can be avoided by ensuring continuity of insulation at all key junctions (roof/wall, wall/window, wall/floor and so on). Refer also to BRE Information Paper IP 1/06, *Assessing the Effects of Thermal Bridging at Junctions and Around Openings*, 2006.

Please refer to Appendix 5.3 of this guide for details.

2.2.4   Ensure that air leakage details comply

Criterion 2: Limits on design flexibility

Air leakage reduces the effectiveness of insulation and other energy conservation measures by up to 40%.

Air permeability is the physical property used to measure the airtightness of the building fabric (paragraph 75 L1A). The design air permeability used in the TER calculation is $10m^3/h \cdot m^2$ @ 50Pa (paragraph 37 L1A); however, it will probably be necessary to design for better performance than this to achieve a satisfactory DER. Some guidance on this can be found in *Limiting Thermal Bridging and Air Leakage: Robust Construction Details for Dwellings and Similar Buildings*, Amendment 1. TSO, 2002.

> Relying on an unusually high standard of airtightness to achieve a DER better than TER also risks relying on an unusually high standard of workmanship, which will be tested at completion.

A new publication (*SEDA Design Guide for Scotland No. 2*, March 2006) provides a lot of useful information on the subject.

Guidance on appropriate air permeability standards for different ventilation strategies can be found in GPG 268 *Energy Efficient Ventilation in Housing*.[11] Higher standards of airtightness are required for whole-house mechanical ventilation systems than for local extract systems with background ventilation.

Please refer to Appendix 6.8 of this guide.

Notes

10. BRE Report 262 *Thermal Insulation: Avoiding Risks*, 2002 edn.

11. GPG 268 *Energy Efficient Ventilation in Housing. A Guide for Specifiers on Requirements and Options for Ventilation*, Revision as CE124/GPG268.

2.2.5   Specify efficient boilers, pipework and controls

Criterion 2: Thermal performance of fabric and services

Boiler efficiency is not a constant and varies according to the season of the year and the varying loads placed on the boiler.

A standard system of tests appropriate to the UK has been devised and is known as the SEDBUK (seasonal efficiency of domestic boilers in the UK) method. All new gas and oil boilers are rated by independently certified testing and placed on the boiler efficiency database at www.sedbuk.com. The list is updated monthly.

The default efficiency for a boiler for the purpose of calculating the TER is a SEDBUK 78% room-sealed boiler with fan flue. In order to meet DER, it is certain that a better performance than this will be required – the *Domestic Heating Compliance Guide*, Table 1, requires 86%.[12] The *Domestic Heating Compliance Guide* is a second-tier document that sets out the recommended standards to meet the requirements of Parts L1A and L1B.

To meet the carbon dioxide emission targets, it will help if primary pipe runs are kept to a minimum, circuits are fully pumped and pipes are well insulated. Controls such as boiler interlock, thermostatic radiator valves and cylinder and room thermostats should limit the time the boiler is running to the minimum, and this is reflected in the SAP calculation.

Please refer to Appendices 6.9, 6.10 and 6.11 for details.

2.2.6   Specify energy-efficient lighting

Criterion 2: Thermal performance of fabric and services

It will generally be adequate to provide a number of fittings that accept only energy-efficient light bulbs (see Appendix 6.14). This will have the effect of limiting the luminaire efficacy (nlum) to 40 luminaire-lumens/ circuit-watt.

The number of energy fittings in the area of the building work should be more than the higher result of either:

- the rate of one per 25m^2 of dwelling (excluding garages); or
- more than one per four fixed light fittings.

(Storage and other areas described in GIL 20 *Low Energy Domestic Lighting* are excluded from this requirement.)

External lighting controlled from the dwelling can be specified in two ways. Lighting fittings can accept either energy-efficient lamps only (efficacy greater than 40 lumens/circuit-watt) or effectively switched lighting (with controls to limit use to only when required at night) using lamps of less than 150W.

Notes

12. *Domestic Heating Compliance Guide*, NBS, 2006.

The notes below paragraphs 42, 43 and 45 of L1A describe in non-technical terms those types of light fitting that are likely to meet these requirements in various circumstances.

2.2.7 Use of mechanical ventilation

Criterion 2: Thermal performance of fabric and services

The performance criteria for mechanical ventilation are described in Part F. The factors to be considered in Part L are the airtightness of the building and duct work, and the efficiency of fans. Natural ventilation is preferred, but external pollution and noise may preclude this. Simple extract fans operated by humidistat or timers to keep running time to a minimum may be needed under Part F to remove condensation from kitchens and bathrooms as well as odours from toilets. Part L requires that these or any other ventilation system should not be overdesigned and should be easy to control and maintain to keep it working efficiently.

Please refer to Appendix 6.12 for details.

2.2.8 Confirm that design is not subject to summer overheating

Criterion 3: Provision of appropriate passive controls to limit solar gains

As increased insulation has reduced heat demand, paradoxically the risk of overheating has increased. Solar gain is the primary cause of external heat gain, but many causes are internal, e.g. from cooking and hot water systems, appliances, lighting and people.

In dwellings, the requirement is that provision should be made to prevent high internal temperature due to solar gain (paragraph 46 L1A). Solar gain is reduced by a combination of choosing an appropriate window size, orientation and shading and by using high thermal capacity construction that can be cooled using night ventilation.

Please refer to Appendix 6.7 for details.

2.2.9   Work out your actual emissions (dwelling carbon dioxide emissions rate, DER)

Criterion 1: Carbon emissions

The above information should allow readers to calculate the actual dwelling carbon dioxide emissions. Two calculations may be required for this, using the same calculation tool as for TER: one optional preliminary version based on plans and specifications as part of the design submission to building control to identify the design features critical to energy performance (so that they can be monitored during construction); the other when the building is complete and air permeability test results are known.

If you do not know the precise detail of, for example, a window, use the default values for the type of window provided in the SAP software.

Appendix B of Part L1A 'Important design features' lists fabric or services that exceed the standards in the AD, and should, therefore, be reported to the Building Control body to demonstrate that the exceptional performance claimed is both correct and achievable, and because they are likely to be necessary to achieve the TER.

Should the designer wish to use any energy-saving technology or solutions that are not included in the published SAP specification, the data must be available on the SAP 2005 website (or a site linked to it). These should be included in the 'Important design features' list under the heading 'Any item involving the application of Appendix Q of SAP 2005'.[13]

Please refer to Appendix 6.2 for details.

2.2.10   Ensure that DER is equal to or better than TER

Criterion 1: Carbon emissions

A note in paragraph 37 of L1A suggests that, in order to achieve the TER, it may be necessary to achieve a design air permeability better than the limit value ($10m^3/h\cdot m^2$ @ 50Pa), particularly when using balanced systems with heat recovery. It is also suggested that U-values will have to be significantly better than those limiting values listed in Part L1A in order to meet the requirement (paragraph 33).

Notes   13. Appendix Q of SAP 2005.

> The process of achieving the target is likely to be iterative. It would be useful to identify those items that can be modified in the SAP model fairly easily with little cost or design impact on the building itself, e.g. provide a more efficient boiler, revise detailing to improve airtightness or improve U-values in roofs and walls.

Once air permeability test results are available, DER calculations should be carried out using the actual test results, together with the performance criteria of all as-built elements, fittings and services. Please refer to 2.2.12 below.

Please refer to Appendix 6.4 for details.

2.2.11   Prescribe a system for site checking by an appropriate person

Criterion 4: As-built performance checks

An example of a compliance checklist is included in Appendix A of Part L1A. However, it is sensible to review the requirements and ensure that ongoing site checks by the appropriate person take place. This is in order to inspect details that affect air permeability, airtightness of ducting, quality of insulation installation and potential thermal bridges (condensation risks) before they are covered by subsequent construction.

Please refer to Appendix 6.16.

2.2.12   Undertake pressure testing during construction

Criterion 4: As-built performance checks

As-built performance checks are likely to fail if the designer has not already taken air permeability into account in the design of the building (see 2.2.4 above).

As approved or accredited details have not yet been published, it is assumed that the provisions of paragraphs 58 and 59 apply. The required pressure tests are to BS EN 13829:2001 Thermal performance of buildings. Determination of air permeability of buildings. Fan pressurisation method. Refer to *CIBSE Technical Memoranda TM23:2000 Testing Buildings for Air Leakage.*

Table 4 of L1A sets out the numbers of tests required. Also note that, if there are fewer than two dwellings on site, it is an option to avoid testing if either the same dwelling type was constructed in the previous 12 months by the same builder and achieved the design air permeability or, alternatively, a default value of 15m³/h·m² is used in the DER calculations (paragraph 63).

Note that half the total number of tests required should be carried out on the first quarter of the dwellings to be completed. This is intended to provide an early indication of any corrective work needed.

Please refer to Appendix 6.8 for details.

2.2.13 Prescribe remedial measures where testing fails

Criterion 4: As-built performance checks

There are transition arrangements in place if air permeability tests carried out before 31 October 2007 fail. This is primarily to give the construction industry time to acquire the necessary knowledge.

Paragraph 61 explains that, before this date, there are two requirements:

First, it will be sufficient to show an improvement in air test results:

- to achieve a result of within 15% of the design air permeability; or
- an improvement of 75% of the difference between the design air permeability and the initial test result (and there is an example of how to do this at the end of paragraph 63 in L1A),

whichever is the easier to achieve.

And second, revise the TER calculation using the air permeability achieved on the initial test results and demonstrate that the DER is no worse than this new TER.

> Although this will pass Part L, the poor airtightness result will appear on the Energy Performance Certificate, which a vendor is required to provide to a purchaser, and this may affect the value of the building.

After 31 October 2007, it will be necessary to demonstrate that the air permeability is not greater than 10m³/h·m² and the DER using the as-built air permeability is still equal to or better than the TER. Remedial works may be required. Details requiring attention can be located using a smoke generator.

Please refer to Appendix 6.8 for further details.

2.2.14 Prescribe commissioning procedures and certification

Criterion 4: As-built performance checks

For construction of a new dwelling, extension of a dwelling, installation of a new system and modification to existing systems, the provisions are identical.

The *heating and hot water system(s)* should be commissioned so that, at completion, the system(s) and their controls are left in working order and can be operated efficiently for the purpose of the conservation of fuel and power.

Schedule 2A in L1A lists self-certification schemes and the type of person qualified to self-certify in these circumstances.

2.2.15 Provide an instruction manual for the heating and ventilation systems

Criterion 5: Provision of guidance for occupiers

Information must be provided to each dwelling 'owner' as required by L1A, paragraph 67. However, paragraph 69 says that the instructions are for the 'occupier'. For the avoidance of doubt, it may be sensible to provide both the owner and the occupier with appropriate instruction manuals. The requirement is for 'sufficient information, including operating and maintenance instructions, enabling the building and the building services to be operated and maintained in such a manner as to consume no more fuel and power than is reasonable in the circumstances'.

Please refer to Appendix 6.18 for details.

Part L1B: existing dwellings 3

Approved Document L1B Conservation of Fuel and Power in Existing Dwellings covers extensions[14] (including **conservatories**) and **material alterations** to existing dwellings, as well as **dwellings** created by a **material change of use**. It also applies to work on certain services or fittings (called **controlled services or fittings**) such as new external doors or windows, new hot water systems, mechanical ventilation and cooling systems, insulation of existing pipes and ducts, and lighting or the provision of a new or changed **thermal element** (wall, roof or floor).

3.1 The approaches and requirements vary with the nature of the work

The requirements vary and depend on whether you are providing an extension to an existing dwelling, adding a conservatory, changing the use of the building or making material alterations or whether you are modifying an existing thermal element or a service. Table 2 shows the requirements for each of the possible approaches.

The general rules of thumb for walls, floors and roofs are as follows:
- where there is a new construction to form an extension or more than 25% of a thermal element is being renovated, the requirements of L1B for *new* thermal elements apply (paragraph 50);
- where an element or fitting is being replaced, the requirements of L1B for *replacement* elements applies (paragraph 51)
- where existing construction is being *renovated* or a conservatory over 30m² is added, the requirements of L1B for *renovated* elements apply (paragraph 54);
- where there is a material change of use or a material alteration, the requirements of L1B for *retained* thermal elements apply (paragraph 56).

Notes

14. There appears to be no explicit upper or lower limit on the size of an extension to a dwelling which would make it a new building (if very large) or exempt (if very small).

3.2   Compliance for work to existing dwellings

The broad requirements and the different methods of meeting them are listed in Table 2.

Table 2
The broad requirements for existing dwellings and how to meet them

Extension – There are three ways to comply with Part L1B if building an extension: options 1, 2 or 3 (paragraph 14)	
Brief definition	An extension has an element of new-build construction required to enlarge an existing dwelling
Option 1	Areas of openings comply (paragraph 15) AND U-values of openings (**controlled fittings**) comply with Table 2(a) (paragraph 31) AND Heating, hot water, pipes, mechanical ventilation, cooling, fixed internal and external lighting (**controlled services**) comply (paragraphs 35–49) AND New thermal elements to Table 3(a) for new elements in an extension (paragraphs 50 and 51) AND As few as possible thermal bridges (paragraphs 52 and 53) AND Reduction of unwanted air leakage (paragraphs 52 and 53) AND Existing opaque fabric that becomes a thermal element (paragraphs 54–57) Subject to **simple payback** calculation
Option 2	Area-weighted U-value complies (paragraph 18b) AND U-values are no worse than Table 1(b) (paragraph 18c)
Option 3	Show CO_2 emission for the actual dwelling plus the actual extension is better than the actual dwelling plus a notional extension using SAP 2005 (paragraph 20)
Conservatories with an area greater than 30m² (paragraph 22) – including substantially glazed extensions	
Brief definitions	A 'conservatory' is an extension with not less than 75% of its roof and not less than 50% of its walls constructed of translucent material and is thermally separated from the main building A 'substantially glazed extension' can have less glazing, but in other respects is the same as a conservatory

Table 2
Continued

Conservatories only	Maintain thermal performance of the wall between the dwelling and the conservatory (paragraph 22a) AND U-values of translucent surface to comply with Table 2(b) as for an existing dwelling (paragraph 22c) AND Independent temperature and on/off controls to heating (paragraph 22b) AND Any heating system to comply with paragraph 35 (paragraph 22b) AND New thermal elements to Table 4(b) as for an existing dwelling (paragraph 22c)
Glazed extensions	Demonstrate that the performance is no worse than a conservatory of the same size and shape (paragraph 24) Area-weighted U-value of the elements is no greater than that of a conservatory as described above (paragraph 24)

Material changes of use to become a dwelling or changes of energy status – There are two ways to comply: options 1 and 2 (paragraph 25)

Brief definition	This applies in this case when a building or a part of a building (even a room) becomes a dwelling. Also, if the number of dwellings changes or the number of rooms used as dwellings changes
Option 1	IMPORTANT – paragraph 61, L1B, requires that an energy rating certificate is provided when there is a material change of use to create a dwelling. This needs to be prepared using a SAP calculation (paragraph 62), i.e. using Option 2 described below. It is not clear under what circumstances the use of Option 1 would be appropriate except as required for comparison with the SAP result (paragraph 28), but it is listed for completeness U-value of existing openings including roof window or rooflight (**controlled fittings**) less than 3.3W/m² to be replaced to Table 2(b) (paragraphs 27e and 32–34) AND Heating, hot water, pipes, mechanical ventilation, cooling, fixed internal and external lighting (**controlled services**) comply (paragraphs 35–48) AND New thermal elements to Table 4(a) (paragraphs 49–53) AND As few as possible thermal bridges (paragraphs 52 and 53) AND Reduction of unwanted air leakage (paragraphs 52 and 53) AND Renovated thermal elements to Table 5(b) (paragraphs 54 and 55) AND Thermal element to be retained and, if worse than Table 5(a), upgraded (paragraphs 56 and 57) subject to **simple payback** calculations (paragraph 72)

Option 2	Calculate whole building CO_2 emission using SAP 2005 AND U-value of individual elements to be no worse than Table 1(b)

Material alterations (paragraph 29)

Brief definition	A material alteration arises when at any stage as a consequence of carrying out building works a **building** or **controlled service** or **fitting** no longer complies with the relevant requirements of Part A (structure, changes to fire safety measures under Parts B1, B3, B4 and B5) or Part M (changes to access and use of buildings) If the building, controlled service or fitting did not comply with these relevant requirements in the first place, a material alteration arises if they become even more unsatisfactory in relation to these requirements When carrying out a **material alteration**, it will be necessary to comply with the following provisions of Part L1B
Reasonable provision	U-value of openings (**controlled fittings**) to Table 2(b) (paragraphs 32–34) AND Heating, hot water, pipes, mechanical ventilation, cooling, fixed internal and external lighting (**controlled services**) comply (paragraphs 35–48) AND New thermal elements to Table 4(a) (paragraphs 54 and 55) AND No individual element worse than Table 4(b) (paragraph 50) AND As few as possible thermal bridges (paragraphs 52 and 53) AND Reduction of unwanted air leakage (paragraphs 52 and 53) (paragraph 45) AND Renovated thermal elements to Table 5(b) (paragraphs 54 and 55) AND Replacement of existing element to Table 4(b) and no worse than Table 1(b) (paragraph 51) AND Elements which become part of the thermal envelope are to be upgraded subject to **simple payback** calculations (paragraphs 56, 57 and 72)

Changes to controlled fittings (paragraphs 32–34)

Brief definition	**Controlled fittings** are windows, roof windows, rooflights and doors
Reasonable provision	U-value of openings (**controlled fittings**) to Table 2(b) (paragraph 32)

Changes to controlled services (paragraphs 35–48)

Brief definition	**Controlled services** are heating and hot water systems, pipes and ducts, mechanical ventilation or cooling, fixed internal and dwelling-supplied external lighting
Reasonable provision	Heating, hot water, pipes, mechanical ventilation, cooling, fixed internal and external lighting (**controlled services**) comply (paragraphs 35–48)

Table 2
Continued

Changes to thermal element (paragraph 49)	
Brief definition	A thermal element is a wall, floor or roof that separates internal conditioned space from the external environment (paragraph 73)
New, in an extension	New thermal elements to Table 4(a) and no individual element worse than Table 1(b) (paragraph 50)
Replacements	Replacements of existing elements to Table 4(b); no individual element worse than Table 1(b) (paragraph 51) AND As few as possible thermal bridges (paragraphs 52 and 53) AND Reduction of unwanted air leakage (paragraphs 52 and 53) (paragraph 45)
Renovated, change which amends thermal performance such as adding new insulation layer	Renovation of more than 25%, then thermal elements to Table 5(b) (paragraph 54) OR (if the above is not technically feasible) Upgrade to the best standard using Table A1 (Appendix of L1B)
Upgraded, where there is a material change of use (to become a dwelling) or an existing non-thermal element becomes a thermal element	Thermal element to be retained is to be upgraded to (paragraphs 56 and 57) subject to **simple payback** calculations (Appendix A of L1B and paragraph 72)

Note that special considerations (relaxations) apply to **historic buildings**, including listed buildings, buildings that are referred to in the local authority's development plan as being of local architectural or historic interest and those buildings in national parks, areas of outstanding natural beauty and world heritage sites (paragraph 8). When dealing with an historic building, reference should be made to the English Heritage Building Regulations and Historic Buildings: *Balancing the Needs of Energy Conservation with those of Building Conservation: An Interim Guidance note on the Application of Part L; Draft Document MK1* (May 2002).

Which road map to follow depends on the specific nature of the work to be done, and this is best resolved using either Table 2 or the flowchart shown in Figure 2. When it is known which requirements apply, a brief explanation in the following text can be found or, if required, there is a more detailed explanation in the appendices.

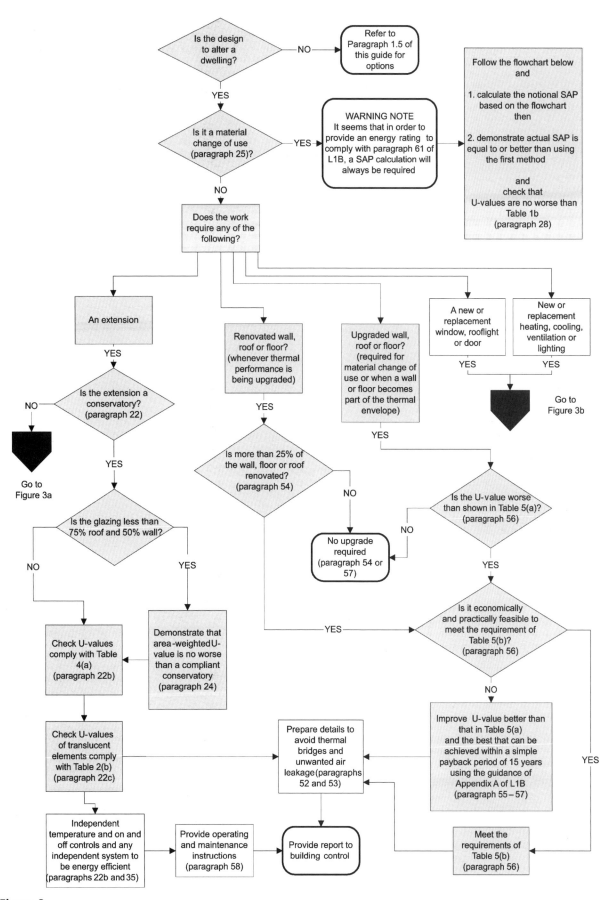

Figure 2
L1B existing dwelling

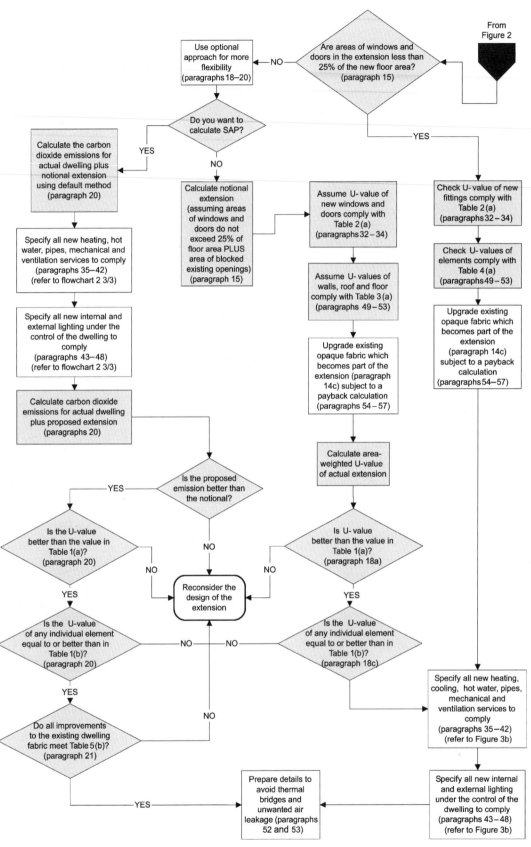

Figure 3a
Existing dwelling (extension)

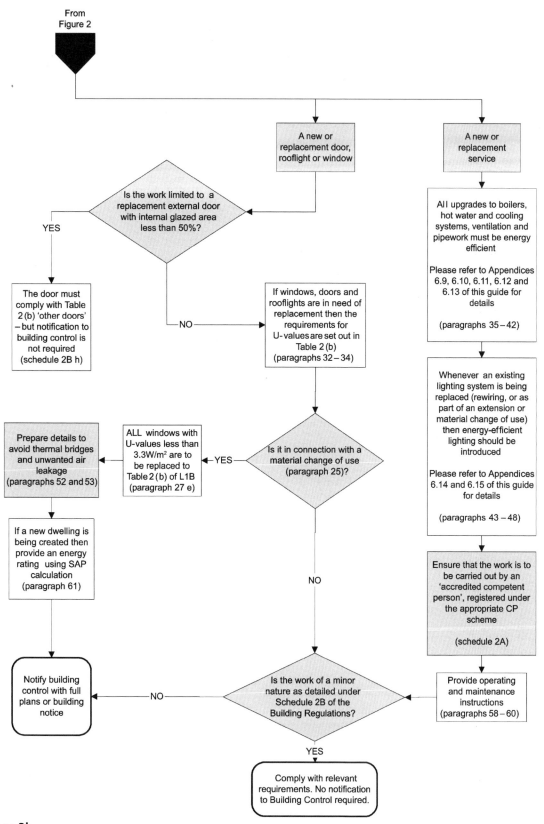

Figure 3b
Existing dwelling (controlled fittings and services)

Note that increasing insulation or introducing other energy efficiency measures into an existing building can increase the risk of condensation forming because of either thermal bridging or interstitial condensation where this did not occur before the works. Paragraph 7 refers to BR 262: *Thermal Insulation: Avoiding Risks* for guidance.

3.2.1 Ensure U-values and areas of openings comply

Applies to extensions (Table 2, option 1) or **material change of use** (Table 2, option 1, requirement for existing windows).

This is the simplest but least flexible approach to achieving the required thermal performance when designing an extension (paragraphs 14, 15, 25–27, 32, 50 and 51 in Part L1B). The U-values of windows and doors should each meet one of the requirements of Table 2(a) in Part L1B, and the walls and roof likewise should meet the requirements of Table 3(a) in Part L1B. The total area of openings is limited to 25% of the floor area *plus* the areas of openings blocked up as a result of extension works. However, if, in complying with this, adequate daylight is not available to BS 8206-2: Code of Practice for Daylighting, it may be possible to adopt a different approach by agreement with the Building Control body.

If it is a **material change of use**, all windows with a U-value of less than 3.3W/m^{2}K should be replaced to the standard shown in Table 2(b) in Part L1B. Walls, roofs and floors should achieve the performance shown in Table 4(b) of Part L1B. A lower standard for fittings and elements than that shown in these tables is acceptable if supported by **simple payback** calculations (see Appendix 6.29 of this guide). However, even then, the values for elements should not be worse than the threshold values shown in Table 5(a) of Part L1B.

3.2.2 Show compliance using the area-weighted U-value calculation

Applies as an alternative approach to 3.2.1 above if more flexibility is required when designing extensions (Table 2, option 2).

There are two ways that U-values can be assessed. Each element in a particular plane has a U-value, and there is the average U-value for all elements of the same type (the area-weighted average U-value). The Regulations allow some (limited) flexibility on U-values calculated in either way, so that in several areas some elements may fall below the default standard assumed in the calculations if they are balanced by enhancements elsewhere; however, no 'individual' U-value should fall below the values shown in Table 1 of Part L1B.

> An 'individual element' is defined in Part L1B following paragraph 18, but is more readily understood as the aggregate of every area that has the same construction regardless of which plane it is in. (This is not the same definition for the U-value of an element, which is normally limited to the same construction in the same plane.) Individual windows are assessed as whole units.

Please refer to Appendix 6.7 for details.

3.2.3 Confirm that thermal bridges are compliant and air leakage is reduced

Applies in all cases where the thermal performance of an element or fitting has been improved.

At the internal surfaces, where insulation is discontinuous or is insufficient (window frames and reveals, at the eaves and so on), the effectiveness of the insulation provided is reduced and cool spots are created, where the local inside surface temperature falls below the dewpoint, causing condensation to form on the inner surface. Guidance can be found on thermal bridging and air permeability in *Limiting Thermal Bridging and Air Leakage: Robust Construction Details for Dwellings and Similar Buildings*, Amendment 1, TSO, 2002.

BR 262[15] shows how thermal bridging can be avoided by ensuring continuity of insulation at all key junctions. BRE IP 1/06[16] is also referred to in Part L, and this provides guidance on thermal bridges such as those found at junctions of floor and roof as well as details around window and door openings.

Please refer to Appendices 6.2 and 6.8 for details.

3.2.4 Show compliance using SAP modelling

Applies as another alternative approach if even more design flexibility is required for extensions (option 3) or for a material change of use (option 2).

For extensions, a more flexible approach is to calculate the carbon dioxide emissions using SAP for the dwelling as it is plus a notional extension (calculated in the same way as 3.2.1 above). Repeat the SAP calculation but using the proposed actual elements and fittings, ensuring that the U-values for each element type are no worse than those shown in Table 1(a) and no worse for each individual element than those shown in Table 1(b).

Notes

15. BRE Report 262 *Thermal Insulation: Avoiding Risks*, 2002 edn.

16. Information Paper IP 1/06 *Assessing the Effects of Thermal Bridging at Junctions and Around Openings in the External Elements of Buildings*, 2006.

(See helpful note in 3.2.2 about 'individual elements'.) The calculation should show that the actual dwelling plus the actual extension has carbon dioxide emissions that are equal to or lower than the actual dwelling plus the notional extension.

For a **material change of use**, a more flexible approach is first to calculate the carbon dioxide emission using the SAP and assuming default values (used for the method described in 3.2.1 above), and then repeat the calculation using the proposed actual performance of the fittings, services and elements. The calculation should show that the proposed actual dwelling has carbon dioxide emissions that are equal to or lower than the notional dwelling.

Please refer to Appendix 6.4 for details on SAP calculations in general.

3.2.5 Justify reduced standards using the 15 year payback criteria

Applies to existing elements that are to be retained but that need to be either renovated or upgraded because they are part of an extension (option 1) or because they are part of a **material change of use** or **change in energy status** (option 1) or a **material alteration**.

> Payback is applicable only when upgrading an existing element or renovating more than 25% of an existing thermal element. Curiously, in the latter case, if less than 25% of an element is to be renovated 'reasonable provision could be to do nothing to improve energy performance' (paragraph 54).

When extending a building OR renewing OR refurbishing a thermal element (wall, roof or floor), there may be situations when full compliance with an upgrade of the U-value is not viable because of economic, functional or technical difficulties. Table A1 of Part L1B gives advice on what is likely to be an economic improvement in U-value when there is little or no existing thermal insulation material present.

Please refer to Appendix 6.29 for details on how to calculate simple payback.

3.2.6 Confirm conservatories (or substantially glazed extensions) comply with requirements

Applies to conservatories or substantially glazed extensions greater than 30m².

A conservatory is defined in Clause 66 of L1B as an extension that is thermally separated from the dwelling by walls, windows and doors with the same U-value (and draught stripping) as the rest of the building. It must also have not less than 75% of the roof area and not less than 50%

of the wall area constructed from translucent material. (Conservatories of less than 30m² are not controlled by the Building Regulations and are exempt.)

The requirements for conservatories are defined in paragraph 22 and are straightforward. According to Table 4(b) of Part L1B, opaque walls should have a U-value better than 0.35W/m²K, whereas floors and the opaque part of the roof (likely to have integral insulation) should have a U-value better than 0.25W/m²K. However, the thermal performance of the floor could be relaxed if the required levels of insulation create problems with floor levels (see footnote to Table 4). The translucent elements should comply with the requirements of Table 2(b) of Part L1B.

A reasonable provision would be for the heating in the conservatory to have separate temperature and on/off controls, but any new heating appliance would have to meet the requirements of paragraph 35, which are discussed in more detail in 3.2.7 below.

Finally, the dwelling should be separated from the conservatory as if the conservatory was an external space, so the walls, doors and windows should be insulated and weather stripped to at least the same standard as the existing dwelling.

3.2.7   Specify efficient boilers, pipework and controls

Applies when new or replacement services are being installed.

Please refer to Appendix 6.9 of this guide for information about efficient boilers. Paragraph 35a.i. refers to the *Domestic Heating Compliance Guide*, which states that levels of performance for new and existing dwellings differ only where practical constraints arise in existing dwellings.

Under certain circumstances (defined in Appendix 1 of the *Domestic Heating Compliance Guide*), entirely new gas-fired boilers can have a SEDBUK value of 78% (for LPG this figure is 80%). Also, the seasonal efficiency of replacement boilers must not be worse than 2% lower than the service being replaced. Gravity feeds should be converted to fully pumped circulation (with by-pass if required by the boiler manufacturer), and the whole system should be properly commissioned. Boilers should be fitted with an interlock control to turn off the boiler when there is no heating or hot water demand. Room thermostats and thermostatic radiator valves should be fitted, but there is some flexibility in the way these can work together.

Zoning is not required when the boiler alone is being replaced. Separate time and temperature controls for the hot water cylinder are not required when only the hot water cylinder is being replaced in an existing system that does not have separate controls.

Accessible pipes should be insulated (labelled as complying with the *Domestic Heating Compliance Guide* or lesser standard if there are practical constraints).

3.2.8   Specify energy-efficient lighting

Required when, as part of an extension, there is a material change of use or rewiring a circuit with new lighting or if new or replacement fixed lighting on the electrical circuit of the dwelling is installed internally or externally.

It will generally be adequate to provide a proportion of fittings that accept only energy-efficient light bulbs. This will have the effect of limiting the luminaire efficacy (nlum) to 40 luminaire-lumens/circuit-watt.

The number of energy fittings in the area of the building work (area of extension, area served or area of newly created dwelling) should be greater than the higher result of either:

- the rate of one per 25m^2 of dwelling (excluding garages); or
- more than one per four fixed light fittings.

(Storage and other areas described in GIL 20 *Low Energy Domestic Lighting* are excluded from this requirement.)

Paragraph 47 allows consideration of the whole dwelling, and the use of ordinary types of lighting in the extension if energy-efficient lighting is introduced to replace existing ordinary types of lighting elsewhere.

External lighting under the control of the dwelling can be specified in two ways. Light fittings can accept either energy-efficient lamps only (efficacy greater than 40 lumens/circuit-watt) or automatically switched lighting (with both effective occupants' controls to limit use to only when required at night) using lamps of less than 150W.

Please refer to Appendix 6.14 for details.

3.2.9   Provide an instruction manual for the heating and ventilation systems

Required when there has been a change to services to ensure that the occupier knows how to operate the new services to maximise energy efficiency.

Information must be provided to the dwelling 'owner' as required by Part L1B, paragraph 58. However, paragraph 60 explains that this information is so that the 'occupier' can operate the systems effectively. For the avoidance of doubt, it may be sensible to provide both the owner and the occupier with appropriate instruction manuals.

The instructions should relate to the installed system and should be simple for householders to understand in terms of adjusting timing and temperature control as well as routine maintenance requirements to achieve efficient operation of the system.

Please refer to Appendix 6.18 for details.

Part L2A: new buildings other than dwellings

Approved Document L2A Conservation of Fuel and Power in New Buildings other than Dwellings covers new buildings other than **dwellings** and includes the first **fit-out works** to a building. This part covers heated common areas in residential buildings as well as nursing homes and student accommodation and, in mixed use developments, the commercial or retail space. It also includes extensions of area greater than 100m² and greater than 25% of the existing floor area.

4.1   The five criteria

> Without compromising the other requirements of the Building Regulations all five criteria have to be met.

They are:

1. the predicted rate of carbon dioxide emissions (**building emission rate**) should be less than the **target emission rate**;
2. the thermal performance of the building fabric and the performance of fixed building services should be no worse than the design limits set out in Part L;
3. those parts of the building not provided with comfort cooling systems should have appropriate passive measures to limit solar gain;.
4. as-built performance should be consistent with the predictions;
5. there should be provision for enabling energy-efficient operation of the building.

4.2   A simple road map for compliance

> By following this map the user will gather enough information to complete the checklist that is included in Approved Document L2A, Appendix A. Paragraph 6 in Appendix A states that the CO_2 target is mandatory, but the other checks in the list represent **'reasonable provision'**.
>
> The checklist includes items that are to be produced by 'the builder'; however, in practice, the designer would have to design and specify solutions that meet the requirements in order for the builder to meet the checklist requirements.

4.2.1   Work out the target emissions (TER)

Criterion 1: Achieving an acceptable building CO_2 emission rate (BER)

TER is expressed as the mass of carbon dioxide in kilograms emitted per square metre of floor area per year as the result of the provision of heating, hot water, ventilation and cooling, and lighting, assuming a notional building (described in paragraph 17 of Part L2A) and a selection of 'standardised activities' when assessed by an **approved calculation tool**. 'Standardised activities' have associated default standards for occupancy times and environmental conditions (temperature, illumination, ventilation rate, etc.). The **approved calculation tool** is either the Simplified Building Energy Model (but only if it can model the building in question) or other approved software (as set out in ODPM Circular 03/2006).[17]

TER is the result of a two-stage calculation. The first stage calculates the CO_2 emissions from heating, hot water, cooling, ventilation and lighting in a notional building with default properties (described in paragraph 22 of Part L2A), but of the same class, size, shape and activity properties as the proposed building. From this, the second stage of the calculation determines TER by applying the procedure described in paragraph 23 of Part L2A. (Details of the information required and how this is actually done is included in Appendix 6.1 of this guide.)

The default U-values for calculating the TER are the same as those required by the 2002 Regulations.[18]

There is an opportunity to use low and zero energy carbon energy sources such as solar hot water, photovoltaic power, etc. (such as that described in *Low or Zero Carbon Energy Sources – Strategic Guide* available on the DCLG website at www.dclg.gov.uk); however, the opportunity to trade this off against other requirements is limited (paragraphs 29–34 of Part L2A).

Notes

17. www.dclg.gov.uk
18. www.ncm.bre.co.uk/faqs.jsp

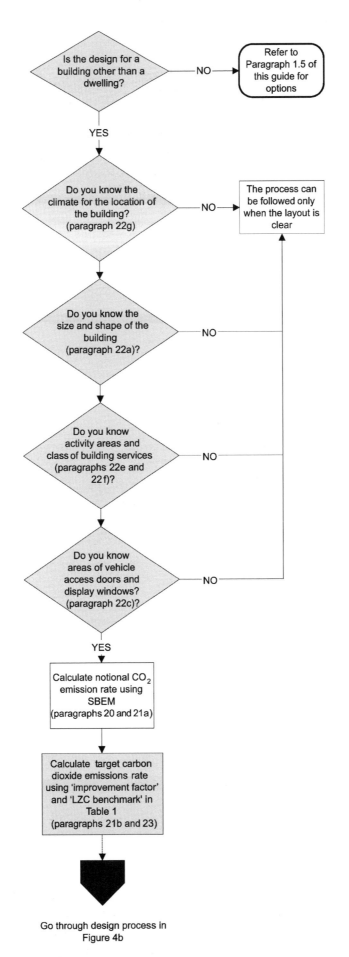

Figure 4a
Road map for meeting the provisions of Part L2A

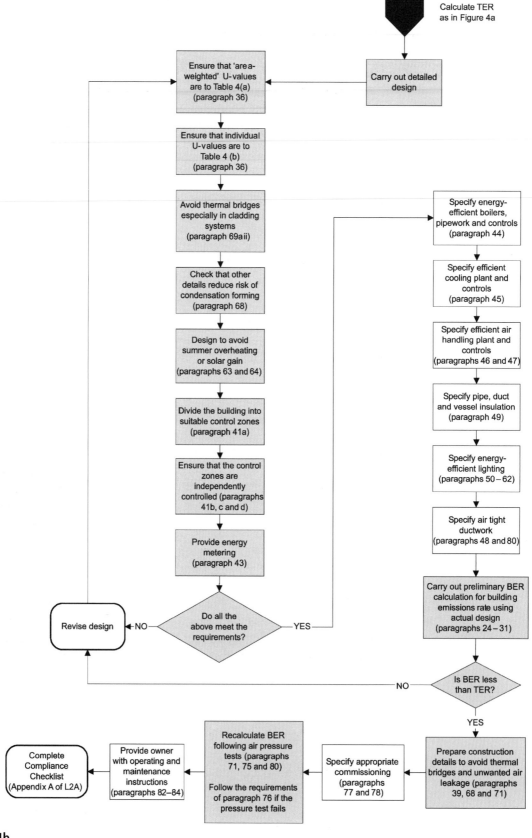

Figure 4b
Road map for meeting the provisions of Part L2A: design process

4.2.2   Ensure U-values comply

Criterion 2: Limits on design flexibility

Table 4 in Part L2A gives U-values for the area-weighted average (column (a)) as well as for the individual element (column (b)).

It is important to check that all U-values in the design meet these minimum levels; however, it is likely that in order to meet the TER these U-values will have to be bettered.

It is also important to check details and calculations to rule out the risk of interstitial condensation (BR 262).[19]

Please refer to Appendix 6.7 for details.

4.2.3   Check for condensation risk in curtain walling and cladding

Criterion 2: Limits on design flexibility

It is important to check details and calculations to rule out the risk of interstitial or surface condensation.

The support structures for curtain walling and cladding systems often form pathways for heat loss through the system. They need to be carefully detailed to avoid thermal bridging as well as to avoid overall worsening of U-value performance. Manufacturers should be able to provide computer-generated calculations to demonstrate adequate performance of key details such as corners, junctions, protrusions and edges and at points of structural support.

Part L refers to *Guidance on the Design of Metal Cladding and Roofing to comply with Approved Document Part L2*, by the Metal Cladding and Roofing Manufacturers Association (MCRMA), which has useful illustrations and explanations of the performance of key details.

Please refer to Appendix 6.6 for details.

4.2.4   Ensure that air leakage details comply

Criterion 2: Limits on design flexibility

Air leakage reduces the effectiveness of insulation and other energy conservation measures by up to 40%.

Air permeability is the physical property used to measure the airtightness of the building fabric (paragraph 71 of Part L2A). The design air permeability used in the TER calculation is $10m^3/h \cdot m^2$ @ 50Pa (Clause 39

19. BRE Report 262 *Thermal Insulation: Avoiding Risks*, 2002 edn.

of Part L2A); however, it will probably be necessary to design for better performance than this to achieve a satisfactory DER, particularly in buildings with mechanical ventilation and air-conditioning.

Please refer to Appendix 6.2.3 of this guide for ways of improving DER. GPG 268 *Energy Efficient Ventilation In Housing*[20] provides guidance on appropriate air permeability standards for different ventilation strategies.

Please refer to Appendix 6.8 for details.

4.2.5   Specify efficient boilers, chillers, pipework, ductwork and controls

Criterion 2: Limits on design flexibility

> There are a number of ways in which Part L encourages the use of energy-efficient equipment without going so far as to insist on them. For example, the use of biomass may require fossil fuel back-up systems; however, in systems producing over 100kW output, the CO_2 emission of the secondary fossil fuel back-up can be ignored (paragraph 24).

Please refer to *The Non-Domestic Heating, Cooling and Ventilation Compliance Guide*. This second-tier document sets out clear guidance on means of complying with the requirements of Part L2A, with sections on boilers, heat pumps, gas- and oil-fired warm air heaters, gas- and oil-fired radiant technology, combined heat and power (CHP), electric space heating, domestic scale hot water, comfort cooling, air distribution systems, pipework and duct insulation.

The guide sets out the minimum provisions for:

- efficiency of the plant that generates heat, hot water or cooling;
- controls to ensure no unnecessary or excessive use of the systems;
- other factors affecting safety or energy efficiency of the system;
- insulation of pipes and ducts;
- acceptable specific fan power ratings for fans serving air distribution systems.

In addition, there is a set of non-prescriptive additional measures to improve plant efficiency.

Notes

20. GPG 268 *Energy Efficient Ventilation in Housing. A Guide for Specifiers on Requirements and Options for Ventilation.* Revision as CE124/GPG268.

The data are provided in a way that is consistent with the national calculation methodology (NCM) used by the Simplified Building Energy Model (SBEM).[21]

Please refer to Appendices 6.9–6.11 for details.

4.2.6   Provide for energy metering

Criterion 2: Limits on design flexibility

Unlike the provisions described in Part L1A, Part L2A acknowledges that properly planned plant control and energy metering equipment can contribute substantially to the energy performance of the building. This means thinking of the building in terms of zones with similar energy control requirements (owing to exposure or use), and giving those zones independent control of timing, temperature and ventilation. Part L2A, paragraph 43, describes a reasonable provision for energy metering that includes assigning energy consumption to use, providing separate meters for monitoring low or zero carbon emission systems and including automatic meter reading and data collection for buildings of area greater than 1000m².

4.2.7   Confirm lighting is suitably efficient

Criterion 2: Limits on design flexibility

The guidance in Part L2A, paragraphs 50–52, relates to lighting for desk-based tasks, and reasonable provision is to provide lighting that on average over the whole area is better than 45 luminaire-lumens/circuit-watt (details of how to calculate this are given in Appendix 6.15).

Paragraph 53 covers general lighting elsewhere, and the requirement is an average 'initial' lamp-plus-ballast efficacy of 50 lamp-lumens/circuit-watt. 'Initial' in this context means the first 100 hours of use.

The requirement for display lighting is initial (100 hour) efficacy of not less than 15 lamp-lumens/circuit-watt in addition to the general lighting requirement (paragraph 60). It should be possible to switch off the display lighting when it is not required.

Notes

21. *The Non-Domestic Heating, Cooling and Ventilation Compliance Guide*, NBS, 2006.

Not subject to Part L are emergency escape lighting and specialist process lighting (such as theatre spotlights, projection equipment, TV and photographic studio lighting, medical lighting in operating theatres and doctors' and dentists' surgeries, illuminated signs, coloured or stroboscopic lighting, and art objects with integral lighting such as sculptures, decorative fountains and chandeliers).

Two units have been used: luminaire-lumens/circuit-watt and lamp-lumens/circuit-watt. A luminaire contains one or more lamps housed in a fitting, and care must be taken to ensure that the correct units are being applied.

It is reasonable provision to locate lighting controls where they would encourage occupiers to switch off lighting when there is sufficient daylight or the space is not in use. Local controls should be no more than 6m away from the luminaire (or a distance of twice the luminaire height from the floor if this is greater). Dimming should be by reduction (and not diversion) of the power supply.

BER can be improved by the use of automatic controls under paragraph 58, but these should be subject to a risk assessment (paragraph 54).

Please refer to Appendix 6.14 for details.

4.2.8 Confirm that the design is not subject to summer overheating

Criterion 3: Limiting the effects of solar gain in summer

Although solar gain is the primary cause of external heat gain, there are also internal causes such as lighting, electronic equipment including computers and, in all cases, human beings.

Paragraph 63 of Part L2A requires that provision should be made to limit solar gain in occupied spaces that do not have comfort cooling (unless the space is next to display glazing, or a stack or unoccupied atrium is used to drive the air circulation).

Please refer to Appendix 6.7 for details.

4.2.9 Work out the building's carbon dioxide emission rate (BER)

Criterion 1: Achieving the TER

Two calculations may be required for BER using the same calculation tool as that used for TER, and one optional preliminary version based on plans and specifications as part of the design submission to the Building Control body. (This is either at plans submission stage or when requested by the Building Control Officer when a Building Notice has been given.) This design stage submission will assist the Building Control body by identifying critical features of the design that will affect the energy performance of the building so that these can be monitored through the construction process. The other submission is made when the building is complete. The second calculation takes into account changes in specification and the actual results of the air permeability, ductwork leakage and fan performance tests.

Please refer to Appendix 6.3 for more details.

4.2.10 Ensure that the BER is equal to or better than TER

Criterion 1: Achieving the TER

This comparison will be generated within the approved modelling software or by using the SBEM. Please see Appendix 6.5 for details of how to generate this comparison through the SBEM.

There is an opportunity in Part L2A to adjust the BER to take advantage of enhanced management and control procedures to improve energy conservation. (See Table 3 for power factor adjustments for enhanced management and control features.)

Please refer to Appendix 6.3 for details.

4.2.11 Prescribe a system for site checking by a qualified inspector

Criterion 4: Quality of construction and commissioning

An example of a compliance checklist is included in Appendix A of Part L2A. However, it is sensible to review the requirements and to ensure that ongoing site checks by the appropriate person take place. This is to allow inspection of the quality of the important energy-saving features of the design to ensure that they are not compromised by workmanship (air permeability, airtightness of ducting, quality of insulation installation and potential thermal bridges and condensation risks) before they are covered by subsequent construction.

Please refer to Appendix 6.16 for more detail.

4.2.12 Undertake fabric leakage and ductwork leakage tests

Criterion 4: Quality of construction and commissioning

There are a number of cases where air testing may not be required, for example:

- An alternative to testing for buildings with a floor area less than 500m² is to assume an air permeability of 15m³/(h·m²); this will mean increasing the performance of the building in other aspects to meet the TER.
- Prefabricated buildings with no site assembly may not require testing, provided routine site-based testing on that module type has been successfully carried out in the past by a third party accredited inspector.
- Large extensions, where sealing off the extension is impractical (this is to be agreed with the Building Control body).
- Large complex buildings where it can be shown and endorsed by a competent person that pressure testing is impractical using the guidance in the ATTMA publication. In this case, an air permeability of not less than 5.0m³/h·m² at 50Pa can be assumed if a suitably qualified person has carried out:
  - a programme of design detail development
  - component testing to show that design air permeability has been achieved and
  - site supervision and
  - a demonstration that a complete air barrier around the whole building envelope will be achieved.
- Buildings that are divided into self-contained units (paragraph 74e), in which case representative tests would be reasonable.

As-built performance checks are likely to fail if the designer has not already taken airtightness of fabric and ducts into account in the design of the building (see 4.2.4 above).

Please refer to Appendix 6.8 for more details.

Ductwork on systems served by fans with a design flow rate greater than 1m³/s (and also ducts designed for BER to have leak rates lower than standard) should be tested for leakage (Part L2A, paragraph 80) in accordance with procedures set out in HVA DW/143 and by a suitably competent person (e.g. a member of the Heating and Ventilation Contractors Association (HVCA) specialist ductwork group or a member of the Association of Ductwork Contractors and Allied Services).

4.2.13   Prescribe remedial measures where testing fails

Criterion 4: Quality of construction and commissioning

Remedial measures will in most cases involve the use of sealants and/or expanding foam fillers unless there are obvious missing draught seals or membranes that can be readily replaced.

If a test result shows that leakage is higher than anticipated, it is recommended that suspect details are selectively blanked off temporarily and the test re-run. This should give valuable information with regard to the design of remedial measures. A list of vulnerable locations is given in Appendix 6.8.

If a building fails to achieve its design air permeability, it may still meet the requirements if it achieves both the target air permeability and building carbon dioxide emission rate. A construction failing on both of these criteria must be retested following the implementation of remedial measures.

If there is a test failure on a representative area of a building compartmentalised into self-contained units as described in paragraph 74e, then after remedial work another representative area should be tested as well.

There are transition arrangements in place for buildings of area less than $1000m^2$ that fail air permeability tests carried out before 31 October 2007. This is primarily to give the construction industry time to acquire the necessary knowledge.

Paragraph 75 explains that, before this date, there are two requirements:

First, it will be sufficient to show an improvement in air test results:

- to achieve a result of within 15% of the design air permeability; or
- an improvement of 75% of the difference between the design air permeability and the initial test result (and there is an example of how to do this at the end of paragraph 76 in Part L2A),

whichever is the easiest to achieve.

Second, revise the TER calculation using the air permeability achieved on the initial test results and demonstrate that the BER is no worse than this new TER.

Although this will pass Part L, the poor airtightness result will appear on the Energy Performance Certificate and may have a negative influence on the value of the building when the owner wishes to sell it.

Please refer to Appendix 6.8 for more details.

If ductwork fails the leakage tests, carry out remedial work and retesting of new sections as set out in DW/143 *A Practical Guide to Leakage Testing*, HVCA, 2000.

4.2.14 Confirm BER is equal to or better than TER as constructed

Criterion 1: Achieving the TER

Once air permeability test results are available, BER calculations should be carried out using the actual test results, together with the performance criteria of all as-built elements, fittings and services. Please also refer to 4.2.13 above.

4.2.15 Prescribe commissioning procedures and certification

Criterion 5: Operating and maintenance instructions

New buildings and works to existing buildings will require commissioning of all controlled services (heating, hot water, electrical and mechanical) in accordance with CIBSE Code M by a suitable person/organisation, which would be a member of the Commissioning Specialists Association or the Commissioning Group of the HVCA.

Note also the requirement for ductwork testing in 4.2.12 above.

Part L2A, Section 3, paragraphs 82–84, requires the preparation of a log book. The log book should be suitable for day-to-day use and (if only by summary reference to operating and maintenance manuals or to the CDM *Health and Safety File*) should include details of any new, renovated or upgraded thermal elements, any new fixed building services, details of their operation and maintenance, any new energy meters and any other details that enable energy consumption to be monitored and controlled.

For guidance and templates, refer to the CIBSE TM31 *Building Logbook Toolkit*, CIBSE, 2006.

Please refer to Appendix 6.17 for details.

4.2.16 Provide an instruction manual for the heating and ventilation systems

The purpose of this is to ensure that the 'owner' and 'occupier' (both terms are used interchangeably in Part L) have all the information needed to run lighting, heating, cooling and ventilation systems effectively and energy efficiently.

An operating and maintenance manual (log book) is to be provided. CIBSE publication TM31 provides guidance on the content, including standard templates. It is recommended in paragraph 84 that an electronic copy is kept of the energy calculation used for TER and DER. The TER and DER should be recorded in the log book.

The log book or equivalent document must be kept up to date with any changes to the building and its systems under Part L2A, paragraph 92.

Designers should ensure that provision of the operating and maintenance manual is included in the contract specification or the mechanical and electrical (M&E) design brief.

Please refer to Appendix 6.18 for details.

Part L2B: existing buildings other than dwellings

5

Approved Document L2B Conservation of Fuel and Power in Existing Buildings other than Dwellings covers **consequential improvements** (required when adding an extension or changing a service to a building with a useful area of more than 1000m²), **extensions** (see note below), **material alterations, material change of use** or **change of energy status** to existing buildings as well as replacement of certain services or fittings (called **controlled services or fittings**) such as new external doors or windows, new hot water systems, mechanical ventilation and cooling systems, insulation of existing pipes and ducts, and lighting, or provision of a new or changed thermal element (wall, roof or floor).

Please note that extensions of more than 100m² AND greater than 25% of the gross area of the existing building are treated as new buildings and Part L2A should be used (although air permeability testing may not be required if it is impractical; refer to 4.2.12 above).

Enclosing an existing courtyard or enclosing the area under an extending roof is regarded as an extension.

Rooms for residential purposes (nursing homes, student accommodation and so on) are not dwellings and so Part L2B applies to these uses.

5.1 The approaches and requirements vary with the nature of the work

The requirements vary and depend on whether the work is improving a building, providing an extension to an existing building, adding a conservatory, changing the use of the building or making material alterations or whether the work is modifying an existing thermal element or a service. Table 3 shows the requirements for each of the possible approaches.

48

5.2 Compliance for work to existing buildings other than dwellings

The road map to follow depends on the specific nature of the work to be done, and this is best resolved using either Table 3 or the flowchart (Figure 5). Once it is known which requirements apply, there is a brief explanation in the following text with a more detailed explanation in the Appendices.

The broad requirements and the different methods of meeting them are listed in Table 3.

Table 3

Consequential improvements (Section 1) (paragraphs 14–23) – There are three reasons to comply: options 1, 2 or 3	
Brief definition	Work that is triggered (as a requirement to the building as a whole) as a consequence of extending an existing building over 1000m² or providing fixed building services or increasing the installed capacity of an existing building service The works which trigger the need for consequential works are called the **principal works**
Option 1 – fabric	Carry out improvements with a **simple payback** of 15 years such as that described in Table 1, L2B (items 1–8), to the extent that their value is less than 10% of the **principal works** The value is to be demonstrated in a report by a suitably qualified person (e.g. a chartered quantity surveyor)
Option 2 – heating service	If the proposal is for initial provision or increased capacity of a *heating* service *per unit area* THEN Improve the performance of the fabric of the part of the building served by the services (paragraphs19a) Upgrade those elements below the threshold U-value in Table 7(a) (paragraphs 22a, 87 and 88) AND Replace existing windows, roof windows or rooflights or doors with U-value worse than 3.3W/m²K (paragraph 22b) (except display windows or high usage entrance doors) AND Check that the improvement is technically, functionally and economically feasible (paragraphs 17, 18 and 107). Work is required only up to the extent that it complies with Part L (Regulation 17D)
Option 3 – cooling	If the proposal is for initial provision or increased capacity of a cooling service per unit area THEN Improve the performance of the fabric of the part of the building served by the services (paragraph 19a)

Table 3
Continued

	Upgrade those elements below the threshold U-value in Table 7(a) (paragraphs 87 and 88)
	AND
	If the windows (excluding display windows) exceed 40% of the façade area and the rooflights exceed 20% of the roof area, and the design solar load exceeds 25W/m²
	THEN (paragraph 23b)
	Upgrade solar controls to achieve:
	a design solar load no greater than 25W/m² OR
	a design solar load reduced by 20% OR
	an effective g-value no worse than 0.3
	AND
	Upgrade the lighting system with an average lamp efficacy of less than 40 lamp-lumens/watt
	AND
	Check that the improvement is technically, functionally and economically feasible (paragraphs 17, 18 and 107). Work is required only up to the extent that it complies with Part L (Regulation 17D)

Extensions (paragraph 24) – may also trigger a **consequential improvement**

Brief definition	An extension has an element of new-build construction required to enlarge an existing building
	(Enclosing a courtyard or enclosing the area under an extending roof is treated as an extension)
	A large extension is not treated as work to an existing building if it has an area greater than 100m², and greater than 25% of the existing floor area. In this case, it is regarded as a new building and part L2A applies (with possible exemption from air permeability testing)
Option 1	Areas of openings comply with Table 2 (paragraph 27)
	AND
	Area-weighted U-values of draught-proofed openings (**controlled fittings**) to comply with Table 5(a) (paragraph 26a)
	AND
	Fixed building services (such as heating, hot water, pipes, mechanical ventilation, cooling, fixed internal and external lighting – **controlled services**) comply (paragraphs 26, 40–74)
	AND
	New thermal elements to Table 6(a) for new elements in an extension (paragraphs 79–84) (paragraph 26b)
	AND
	As few as possible thermal bridges (paragraphs 83 and 84)
	AND
	Reduction of unwanted air leakage (paragraphs 83 and 84)
	AND
	Existing opaque fabric that becomes a thermal element (paragraphs 87 and 88)
	Subject to **simple payback** calculation (paragraphs 26c and 107)

Option 2	Area-weighted U-value is no worse than Table 3(a) (paragraph 29) AND Individual U-value is no worse than Table 3(b) (paragraph 29)
Option 3	Show CO_2 emission for actual building plus extension is better than notional building plus extension using an accredited calculation tool (in the calculation, the notional and actual building should incorporate the actual proposed **consequential improvements**) (paragraph 30) AND U-value of each element type no worse than Table 3(a) (paragraph 30) AND U-value for any individual element no worse than Table 3(b) (paragraph 30) AND Upgrades to existing building no worse than Table 7(b) (paragraph 31)

Conservatories or substantially glazed extensions with an area greater than 30m² (paragraph 32) – may also trigger a consequential improvement

Brief definition	An extension with not less than 75% of its roof and not less than 50% of its walls of translucent material and thermally separated from the main building A 'substantially glazed extension' can have less glazing, but in other respects is the same as a conservatory
Conservatories	Maintain thermal performance of wall between building and conservatory (paragraph 32a) AND U-values of translucent surface to comply with Table 5(b), as for an existing building AND U-values of opaque elements no worse than Table 3(b) (paragraph 32c) Independent temperature and on/off controls to fixed building services (paragraph 32b) AND Any heating service to comply with paragraph 41 (paragraph 32b)
Glazed extensions	Demonstrate that the performance is no worse that a conservatory of the same size and shape (paragraph 33) Area-weighted U-value of the elements is no greater than that of a conservatory as described above (paragraph 33)

Material changes of use or changes of energy status (paragraphs 34 and 35)

Brief definition	This applies in this case (Part L2B) when a building or a part of a building changes to a use other than to a dwelling (If the change of use is to a dwelling, then L1B would apply) Note: It would appear (but it is not obvious) that if the material change of use is greater than 100m², or greater than 25% of the volume of the existing building, then the work should meet the requirements of L2A (paragraph 35 refers to paragraph 25) and **consequential improvements** may be required

Table 3
Continued

Option 1	U-value of existing openings including roof window or rooflight less than 3.3W/m² to be replaced (except display windows and high usage entrance doors) to Table 5(b) (paragraphs 36e and 75–78) AND Fixed building services (such as heating, hot water, pipes, mechanical ventilation, cooling, fixed internal and external lighting – **controlled services**) comply (paragraphs 36a and 40–76) AND New thermal elements to Table 6(a) (paragraphs 36b and 79–84) AND As few as possible thermal bridges (paragraphs 83 and 84) AND Reduction of unwanted air leakage (paragraphs 83 and 84) AND Renovated thermal elements to Table 6(b) (paragraphs 36c, 85 and 86) Thermal element worse than Table 7(a) to be upgraded to Table 7(b) subject to **simple payback** calculation (paragraphs 36d, 87 and 88)
Option 2	Calculate whole-building CO_2 emission using an accredited whole-building calculation model to demonstrate that it will become no worse than following Option 1 (paragraph 37) U-value of any individual element is no worse than Table 3(b) (paragraph 37)
Material alterations (paragraph 38)	
Brief definition	A material alteration arises when at any stage as a consequence of carrying out building works a building or **controlled service** or **fitting** no longer complies with the relevant requirements of Part A (structure, changes to fire safety measures under Parts B1, B3, B4 and B5) or under Part M (changes to access and use of buildings) (If the building, controlled service or fitting did not comply with these relevant requirements in the first place, a material alteration arises if they become even more unsatisfactory in relation to these requirements) When carrying out a **material alteration**, it will be necessary to comply with the following provisions of Part L2B

Existing window and other openings with U-values worse than 3.3W/m^2 to be replaced to comply with Table 5(b) (paragraphs 39c and 75–78)

AND

Fixed building services (such as heating, hot water, pipes, mechanical ventilation, cooling, fixed internal and external lighting – **controlled services**) comply (paragraphs 39d and 41–74)

AND

New thermal elements to Table 6(a) (paragraphs 39a and 79–84)

AND

As few as possible thermal bridges (paragraphs 83 and 84)

AND

Reduction of unwanted air leakage (paragraphs 83 and 84)

AND

Renovated opaque thermal elements to Table 7(b) (paragraphs 39b, 85 and 86)

Any element that becomes part of the thermal envelope is to be upgraded (paragraphs 39c, 87 and 88) subject to **simple payback** calculations

Changes to controlled fittings (paragraphs 75–78)

Brief definition	**Controlled fittings** are windows (including the glazed elements of a curtain wall), rooflights and doors (including large access doors for vehicles and roof ventilators) (but not display windows or high usage doors)
Reasonable provision	Area-weighted U-value of draught-proofed replacement fittings in openings to comply with Table 5(b) (paragraphs 75–78)

Changes to controlled services (paragraphs 41–74) – may also trigger a consequential improvement

Brief definition	**Controlled services** are heating and hot water systems, pipes and ducts, mechanical ventilation or cooling, fixed internal lighting including display lighting and occupier-controlled external lighting (it does not include emergency or specialist process lighting; refer to note in 4.2.7 of this guide for more details)
Reasonable provision	New or upgraded fixed building services (such as heating, hot water, pipes, mechanical ventilation, cooling, fixed internal and external lighting – **controlled services**) comply (paragraphs 42–74) OR For central plant, an efficiency that it is not less than that of the service being replaced (paragraph 41aii)

Table 3
Continued

Changes to thermal element (paragraphs 79–88)	
Brief definition	A thermal element is a wall (including the opaque elements of a curtain wall), floor, ceiling or roof that separates internal conditioned space from the external environment (paragraph 109)
New	New thermal elements to Table 6(a) (paragraphs 79–82) AND As few as possible thermal bridges (paragraphs 83 and 84) AND Reduction of unwanted air leakage (paragraphs 83 and 84)
Replaced	When constructed as a straight replacement of an existing element to Table 6(b) AND Provided no individual element has a U-value worse than Table 3(b)
Renovated	Renovation of more than 25%, then thermal elements to Table 7(b) (paragraphs 85 and 86)
Upgraded	Thermal element to be retained that needs to be upgraded in connection with other works (material change of use or when it changes to a thermal element or is part of a **consequential improvement**) (paragraphs 85–88) subject to **simple payback** calculations

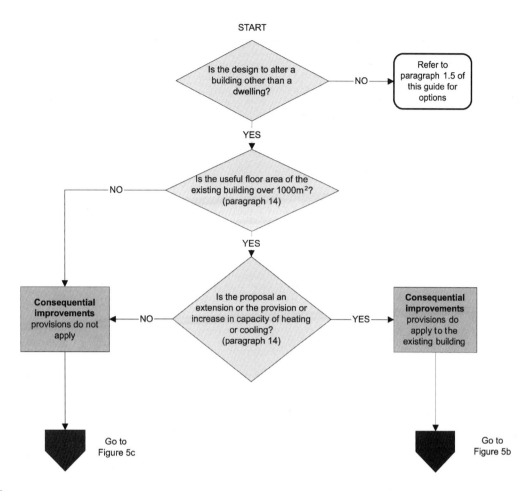

Figure 5a
L2B consequential improvements. Assess the need to provide 'consequential improvements' to the existing building as a result of alterations

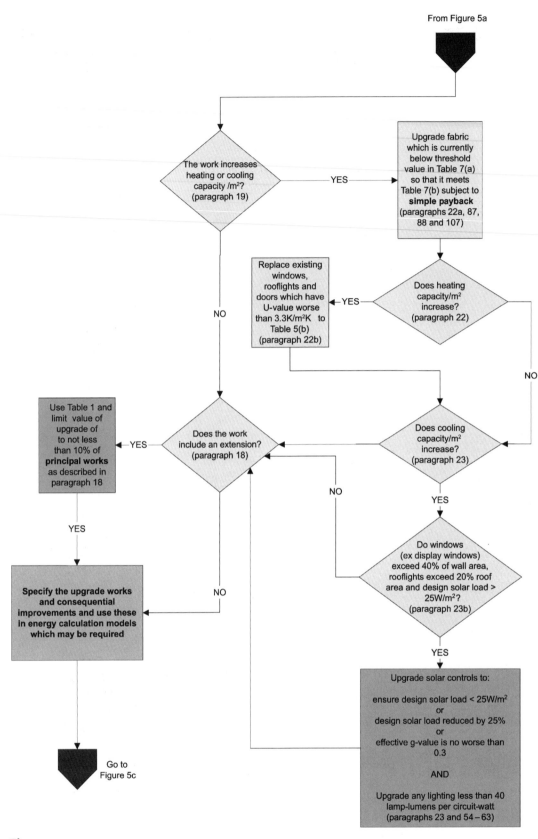

Figure 5b
L2B consequential improvements (cont.). Assess what 'consequential improvements' are needed to meet the requirements

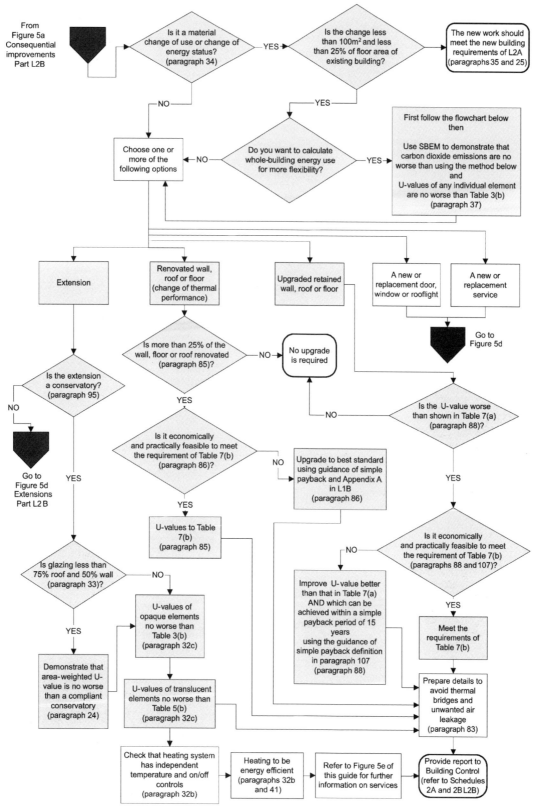

Figure 5c
L2B consequential improvements (cont.). Walls, roofs, floors and conservatories

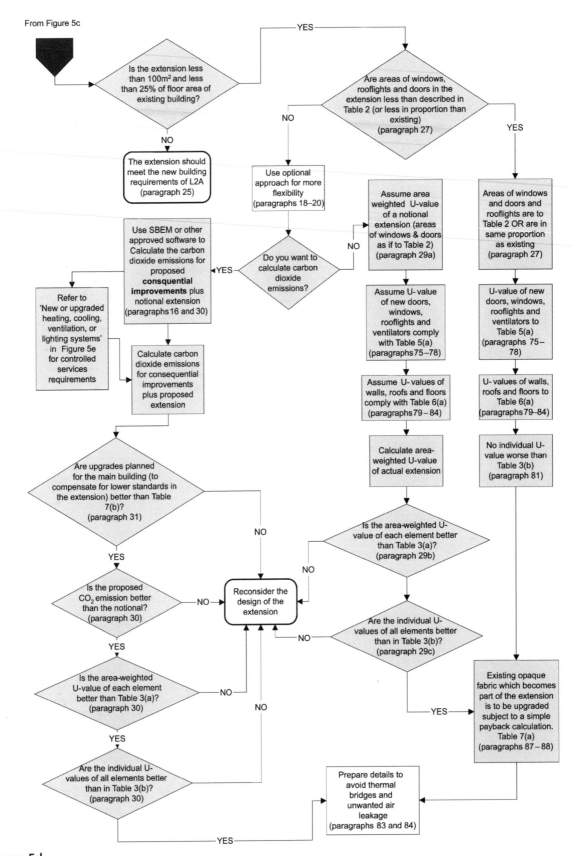

Figure 5d
L2B consequential improvements (cont.). Designing a new extension

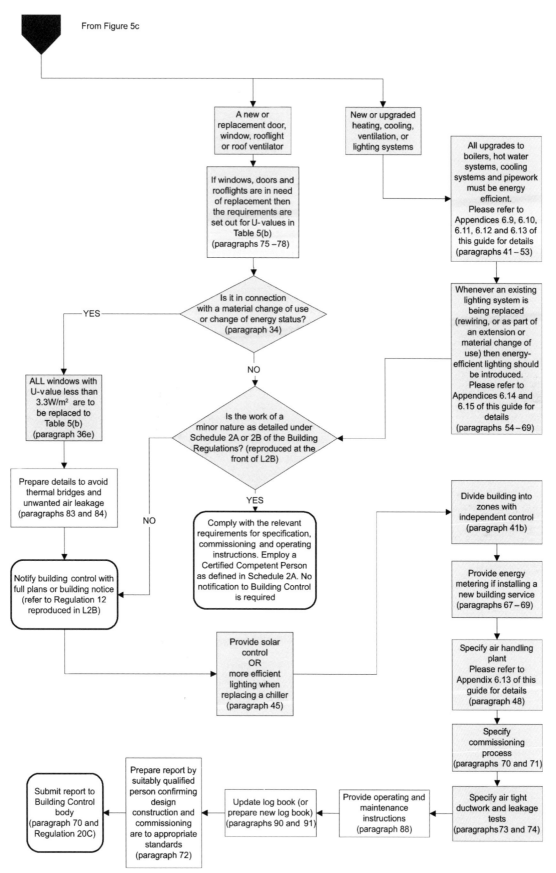

Figure 5e
L2B consequential improvements (cont.). Windows, doors, rooflights and services

5.2.1 Assess the need to include 'consequential improvements': Section 1

In principle, this is a simple provision. If certain defined kinds of work are being carried out to a large building with a useful area over 1000m^2, then there is a requirement to upgrade the energy conservation performance of that building provided the upgrade is economically, functionally and technically feasible. These upgrades are called **consequential improvements**.

The defined kinds of work are building an extension, providing an initial installation of heating and/or cooling, and increasing the capacity of heating and/or cooling per m^2. These are called the **principal works.**

The amount that your client will be required to spend on economically feasible **consequential improvements** is limited by paragraph 18 of Part L2B to **simple payback** in 15 years and by paragraph 20 to a value not exceeding 10% of the value of the **principal works** demonstrated to be the case in a signed report by a competent person (such as a chartered quantity surveyor).

Part L2B, Table 1, lists some guide measures that would be regarded as economically feasible improvements if the building is being extended. Generally, they focus on upgrading old (more than 15 years old, therefore inefficient) services, introducing energy metering and upgrading the performance of the building thermal envelope. However, there is also the opportunity (if the on-site systems produce less than 10% of the energy demand) to introduce zero or low carbon emission energy-generating technology, but the simple payback period has been reduced to 7 years for this technology.

If the building work is for an initial installation or an increase in capacity per m^2 of heating or cooling, the first priority is to improve the insulation of thermal elements (demonstrating that this is technically and economically feasible using simple payback) and reduce air leakage. The second priority is to upgrade glazed elements, to reduce heat loss in the case of heating or reduce heat gain by solar control shading in the case of cooling. An alternative to providing solar control shading is to provide more efficient lighting systems. There is no 10% limit for these improvements; all cost-effective improvements should be carried out. The 10% limit applies only to extensions.

5.2.2 Justify, if necessary, reduced standards using the 15 year payback criterion

Applies to consequential improvements or upgrading existing elements owing to a material change of use or where an existing element becomes part of the thermal envelope. It also applies when a window needs to be improved beyond the minimum U-value of 3.3W/m^{2}K because of a consequential improvement, material change of use or change of energy status.

This provision requires that only economically feasible upgrade work measured using simple payback needs to carried out, so permitting some relaxation in the standards of thermal performance required. However, the minimum expenditure required for upgrading elements and fittings is 10% of the cost of the **principal works** where there are **consequential improvements** unless the existing building can be upgraded to a compliant standard for less than this.

5.2.3   Ensure U-values comply

Applies to thermal elements (such as new or upgraded walls, roofs and floors) and controlled fittings (such as windows, rooflights, roof windows and doors) in any work to existing buildings.

The designer needs to be able to confirm that the design meets the appropriate criteria at design submission and on completion of construction. If the carbon dioxide emissions have been calculated using approved energy modelling software, then checks need to be carried out that the specified fittings and elements meet the assumptions of the model. If the window is specified by performance criteria, and thus U-values cannot be calculated, the performance for a standard configuration from BR 443 or Table 6e in SAP 2005 can be used. SAP is the Government's approved methodology for energy rating of dwellings (*SAP Rating Guide*), which can be found at www.bre.co.uk/sap2005.

Part L2B contains several tables giving U-values for the thermal performance of elements and fittings. For instance, there are seven different U-value references for walls, five for floors, 12 for roofs and more than 18 for glazed elements, so a little care is needed.

Please refer to Appendix 6.7.

> Display windows (defined in Section 5 of Part L2B) do not have to meet U-value standards.

Table 2 in Part L2B shows the area of openings permitted in an extension expressed as a percentage of wall area (or roof area). This can be increased if the percentage of opening in the existing building is already larger than the figures shown in the table.

5.2.4 Show compliance using area-weighted U-value calculation

Extensions, material changes of use and the upgrade of existing controlled fittings. This is an alternative approach to provide more flexibility.

Area-weighted U-values allow some elements to fall in performance below the average required U-value for each element provided it is compensated by improved performance elsewhere. However, U-values should not fall below the limit shown in Table 3 of Part L2B.

There is more detailed explanation in Appendix 6.7.

5.2.5 Show compliance using modelling

Compliance can be shown through modelling either by approved modelling software that follows the NCM or by using the SBEM. Please see Appendix 6.7 for details of how to show compliance through the SBEM.

5.2.6 Confirm thermal bridges comply and that air leakage is reduced

Applies in all cases where thermal performance of an element or fitting has been improved (paragraphs 85 and 86).

At the internal surfaces, where insulation is discontinuous or is insufficient (window frames and reveals, at the eaves and so on) the effectiveness of the insulation provided is reduced and cool spots are created, where the local inside surface temperature falls below the dewpoint, causing condensation to form on the inner surface. Guidance can be found on thermal bridging and air permeability in *Limiting Thermal Bridging and Air Leakage: Robust Construction Details for Dwellings and Similar Buildings*, Amendment 1, TSO, 2002.

BR 262[22] shows how thermal bridging can be avoided by ensuring continuity of insulation at all key junctions. BRE IP 1/06[23] is also referred to in Part L, and this provides guidance on thermal bridges such as those found at junctions of the floor and roof as well as details around window and door openings.

Please refer to Appendices 6.2 and 6.8 for details.

Notes

22. BRE Report 262 *Thermal Insulation: Avoiding Risks,* 2002 edn.

23. Information Paper IP 1/06 *Assessing the Effects of Thermal Bridging at Junctions and Around Openings*, 2006.

5.2.7   Specify efficient boilers, chillers, pipework, ductwork, fans and controls

The Non-Domestic Heating, Cooling and Ventilation Compliance Guide is a second-tier document that sets out clear guidance on means of complying with the requirements of Part L2B with sections on boilers, heat pumps, gas- and oil-fired warm air heaters, gas- and oil-fired radiant technology, combined heat and power (CHP), electric space heating, domestic scale hot water, comfort cooling, air distribution systems, pipework and duct insulation.

The guide sets out the minimum provisions for:

- efficiency of the plant that generates heat, hot water or cooling;
- controls to ensure no unnecessary or excessive use of the systems;
- other factors affecting safety or energy efficiency of the system;
- insulation of pipes and ducts;
- acceptable specific fan power ratings for fans serving air distribution systems.

In addition, there is a set of non-prescriptive additional measures to improve plant efficiency.

If central plant is being replaced, it should meet the requirements of the compliance guide, but even so its efficiency should not be worse than the equipment being replaced (paragraph 43(a.ii) of Part L2B). There are adjustment factors to apply if the fuel source is being changed.

The building should be divided into separately controlled zones (paragraph 43(b.i)) corresponding to areas with different solar exposure, occupancy period or activity. The plant for each zone should operate only when required with the default setting as off. The controls required for the type and size of plant are specified in various tables throughout the compliance guide. Please refer to Appendix 6.10 for details.

Ductwork should be insulated against heat gain/loss (Section 10 and Table 39 of the *Non-Domestic Heating, Cooling and Ventilation Compliance Guide*) and tested in accordance with the procedures of HVCA DW/143.

Cooling loads should be reduced as much as possible by reducing solar and internal heat gains and natural ventilation should be used without cooling plant, but only if comfort conditions can be achieved. Fans should meet the requirements of paragraph 51 and Table 36 of the *Non-Domestic Heating, Cooling and Ventilation Compliance Guide*.

Pipework should be insulated to Section 10 and Table 37 of the *Non-Domestic Heating, Cooling and Ventilation Compliance Guide*.

Please refer to Appendices 6.9–6.13 for details.

5.2.8   Provide for energy metering

The requirement under Part L2B is to demonstrate that reasonable provision of energy meters has been made for effective monitoring of the performance of newly installed plant (paragraph 43(e)). The aim is to enable building occupiers to assign at least 90% of their energy consumption of each fuel to the various end uses (such as heating and lighting). Meters should be installed for the services that form part of the works in accordance with the recommendations of CIBSE TM39. Low and zero carbon (LZC) systems should be separately monitored, and, in buildings with a total useful floor area of greater than 1000m², automatic meter reading and data collection should be provided (paragraphs 69–71).

5.2.9   Confirm lighting is suitably efficient

> Not subject to Part L are emergency escape lighting and specialist process lighting (such as theatre spotlights, projection equipment, TV and photographic studio lighting, medical lighting in operating theatres and doctors' and dentists' surgeries, illuminated signs, coloured or stroboscopic lighting, and art objects with integral lighting such as sculptures, decorative fountains and chandeliers).

Part L2B also applies when new lighting is provided to less than 100m² of space but this does not need to be notified to the Building Control body.

The guidance in Part L2B, paragraphs 57–59, relates to lighting for desk-based tasks, and reasonable provision is to provide lighting that on average over the whole area is better than 45 luminaire-lumens/circuit-watt after taking into account a control factor that can enhance lighting efficiency. This is described in the help note below (see also paragraph 59c and Table 4 of Part L2B).

> Under Part L2B, in existing buildings a control factor can be applied to increase the efficacy of lighting if automatic controls turn off lights when they are not required. This is not an appropriate option in Part L2A, which already takes automatic control into account in the BER calculation.

Details of how to calculate this are in Appendix 6.15.

Paragraph 60 covers general lighting elsewhere, and the requirement is an average 'initial' lamp-plus-ballast efficacy of 50 lamp-lumens/circuit-watt. 'Initial' in this context means the first 100 hours of use.

The requirement for display lighting (paragraph 66) is initial (100 hour) efficacy of not less than 15 lamp-lumens/circuit-watt in addition to the general lighting requirement. The display lighting should be capable of being switched off when not required.

Please refer to Appendix 6.15 for details.

5.2.10 Undertake fabric leakage, ductwork leakage and fan performance testing

> Large extensions that come under the provisions of Part L2A may not require pressure testing owing to the difficulty of containing the extended area. Please refer to Appendix 6.8 for details.

Fabric leakage tests are not required under Part L2B, but ductwork should be tested in accordance with the procedures set out in HVCA DW/143 on systems served by fans with a design flow rate greater than 1m³/s and on certain classes of high-pressure duct described in HVCA DW/143. The tests should be carried out by a suitably qualified person who is a member of the HVCA Specialist Ductwork Group or the Association of Ductwork Contractors and Allied Services.

5.2.11 Prescribe remedial measures when testing fails

If ductwork fails the leakage tests, carry out remedial work and retesting of new sections as set out in DW/143 *A Practical Guide to Ductwork Leakage Testing*, HVCA, 2000.

5.2.12 Prescribe commissioning procedures and certification: log book

Works to existing buildings require commissioning of all controlled services (heating, hot water, electrical and mechanical) in accordance with CIBSE Code M by a suitable person/organisation, e.g. a member of the Commissioning Specialists Association or the Commissioning group of the HVCA.

Note also the requirement for ductwork testing described in 5.2.10 above.

Section 4 of Part L2B, paragraphs 91–94, requires either the update of an existing log book or the preparation of a new one if one does not exist. The log book should be suitable for day-to-day use and, if only by summary reference to operating and maintenance manuals or to the CDM

Health and Safety File, include details of any new, renovated or upgraded thermal elements, any new fixed building services, details of their operation and maintenance, any new energy meters and any other details that enable energy consumption to be monitored and controlled.

For guidance and templates, refer to the CIBSE TM31 *Building Logbook Toolkit*.

Please refer to Appendix 6.17 for details.

5.2.13 Provide an instruction manual for the heating, cooling, ventilation and lighting systems

The purpose of this is to ensure that the 'owner' and 'occupier' (both terms are used interchangeably in Part L) have all the information needed to run lighting, heating, cooling and ventilation systems effectively and energy efficiently.

An operating and maintenance manual (log book) is to be provided. CIBSE publication TM31 provides guidance on the content, including standard templates. It is recommended in paragraph 84 that an electronic copy is kept of the energy calculation used for TER and DER. The TER and DER should be recorded in the log book.

The log book or equivalent document must be kept up to date with any changes to the building and its systems under Part L2B, paragraph 92.

Designers should ensure that provision of the operating and maintenance manual is included in the contract specification or the M&E design brief.

Please refer to Appendix 6.18 for details.

General appendices

6.1 Target carbon dioxide emission rate (TER)

6.1.1 What is the TER?

The TER is the minimum energy performance requirement for new dwellings. TER is expressed as mass of carbon dioxide (CO_2) in kg per m² of floor area that should be emitted per year, as a result of heating, hot water, ventilation, cooling and internal fixed lighting.

6.1.2 Calculate the TER (dwellings)

The TER must be calculated with the same tool as intended for the dwelling carbon dioxide emission rate (DER).

> Part L1A says that these calculations for individual dwellings up to 450m² should be calculated using the SAP[24] and, for larger dwellings, the SBEM[25] (see Appendix 6.5).

The TER is calculated in two stages:

First, calculate the CO_2 emission rates (heating and hot water (C_H) and fixed lighting (C_L)) from a notional dwelling of the same size and shape as the actual dwelling and that is constructed according to the reference values as set out in Appendix R1 of the SAP. This table is reproduced here as Table 4.

Notes

24. The Government's *Standard Assessment Procedure for Energy Rating of Dwellings*, Defra, 2005.

25. SBEM publications.

Table 4
SAP Guide, Appendix R1: reference values (to be used when calculating the CO_2 emissions for a notional dwelling)

Reference factor

Element or system	Value
Size and shape	Same as proposed dwelling
Opening areas (windows and doors)	25% of total floor area (or, if less, the exposed façade area) One opaque door of area 1.85m² Any other doors fully glazed
Walls	U = 0.35W/m²K
Floors	U = 0.25W/m²K
Roofs	U = 0.16W/m²K
Opaque door	U = 2.0W/m²K
Windows and glazed door	U = 2.0W/m²K Double glazed, low E hard coat Frame factor 0.7 Solar energy transmittance 0.72 Light transmittance 0.80
Living area fraction	Same as proposed dwelling
Shading and orientation	All glazing orientated E/W Average overshading
Number of sheltered sides	2
Allowance for thermal bridging	0.11 × total exposed surface area (W/K)
Ventilation system	Natural ventilation with intermittent extract fans
Air permeability	10m³/m²/h at 50Pa
Chimneys	None
Open flues	None
Extract fans	Three for dwellings with floor area greater than 80m² Two for smaller dwellings
Primary heating fuel (space and water)	Mains gas
Heating system	Boiler and radiators Water pump in heated space
Boiler	SEDBUK 78% Room-sealed Fanned flue
Heating system controls	Programmer + thermostat + TRVs Boiler interlock
Hot water system	Stored hot water, heated by boiler Separate time control for space and water heating
Hot water cylinder	150 litre cylinder insulated with 35mm of factory-applied foam
Primary water heating losses	Primary pipework not insulated, cylinder temperature controlled by thermostat
Secondary space heating	10% electric
Low-energy light fittings	30% of fixed outlets

L1A *Work in New Dwellings*, 2006 edition. Approved Document L1A *Conservation of Fuel and Power*.

Then, determine the TER using the following formula:

$$TER = (C_H \times \text{fuel factor} + C_L) \times (1 - \text{improvement factor})$$

The fuel factors are listed in Table 1 in AD L1A,[26] and the improvement factor is currently 20%.

For buildings with more than one dwelling, TER can be calculated using the following formula:

$$\frac{(TER_1 \times \text{floor area}_1) + (TER_2 \times \text{floor area}_2) + (TER_3 \times \text{floor area}_3)}{(\text{floor area}_1 + \text{floor area}_2 + \text{floor area}_3)}$$

6.1.3   Calculate the TER (buildings other than dwellings)

The TER for non-dwellings must be calculated with the same tool as the building carbon dioxide emission rate (BER).

To make these calculations, use the SBEM[27] (see Appendix 6.5) or another approved modelling tool.[28]

The TER is calculated in two stages:

First, calculate the CO_2 emission rate ($C_{notional}$) from a notional building described below.

Then, adjust the first calculation by an improvement factor described below.

6.1.4   Notional building

The notional building required for the first step of the TER calculation:

- is the same size and shape as the proposed building;
- complies with the energy performance values in SBEM and ODPM national calculation methodology (basically using Part L 2002 U-values as defaults);
- has the same area of vehicle access doors and display windows as the proposed building;
- excludes any service not covered by Part L, such as emergency lighting, specialist process lighting and lifts (refer to Appendix 6.14 of this guide for the full list of lighting which is not covered by part L);
- has the same activity areas and class of services as the proposed building using pre-defined SBEM definitions;

Notes

26. Table 1, paragraph 21, Part L1A.
27. SBEM publications.

28. Approved by ODPM as a national calculation methodology tool for demonstrating compliance with the Building Regulations Part L.

- is subject to occupancy times and environmental conditions in each activity area as defined by the standard data associated with reference schedules;
- is subject to the climate defined by the CIBSE test reference year for the site as appropriate to location;
- uses mains gas for heating if available or oil if not; electricity for all other services;
- uses CO_2 emissions factors from Table 2 in Part L2A.

The TER is based on a notional building with modest amounts of glazing. Buildings that allow greater solar gain will have to compensate through enhanced energy efficiency measures in other aspects of the design.

6.1.5   Improvement factors and the low and zero carbon (LZC) benchmark

The improvement factor is the improvement in energy efficiency as given in Table 1 of Part L2A appropriate to the classes of building services in the proposed building. The LZC benchmark is the provision for low and zero carbon energy sources as given in Table 1 in Part L2A.

Further information can be found in *Low or Zero Carbon Energy Sources: Strategic Guide*, NBS, 2006, also available on the ODPM website at http://www.odpm.gov.uk.

Designers can choose to include more renewables than the benchmark provision, although the extent to which this can be traded off against fabric measures is limited. A lesser renewables provision would have to be compensated for by enhanced energy efficiency measures.

6.1.6   The formula

The TER is obtained from the following formula:

$$TER = C_{notional} \times (1 - \text{improvement factor}) \times (1 - \text{LZC benchmark})$$

There is an example of how the fixed values of the formula work in the footnote to Table 1 of Part L2A.

6.2 Calculating DER for dwellings

6.2.1 What is the dwelling carbon dioxide emission rate (DER)?

The DER is the calculated predicted mass of carbon dioxide (CO_2) emitted by a dwelling per m² of floor area per year. It is calculated using the SAP (see Appendix 6.4) at both pre-construction stage (optional) for design submission to building control and for the as-built dwelling at practical completion using the results from the air permeability tests. It is a mandatory requirement that the DER must be better than the TER in order that an energy performance certificate can be issued.

The DER must be calculated with the same tool as the TER.

The calculation, for individual dwellings of floor area no greater than 450m², must be carried out using the standard assessment procedure (SAP 2005) (see Appendix 6.3);[29] in addition, for larger individual dwellings, the SBEM (see Appendix 6.5) should be used.[30]

In the author's opinion, it would be unlikely that requirements for the issue of a practical completion certificate would be met without the issue of the Energy Performance Certificate and, hence, DER of the as-built construction being better than TER.

6.2.2 Criteria for demonstrating compliance

The need to calculate DER and show that it is better than TER (see Appendix 6.2.3) is a central requirement of the Building Regulations.

The key criterion is achieving an acceptable DER – the predicted rate of carbon dioxide emissions from the dwelling (the DER) is not greater than the target rate (the TER). Within this requirement, there is scope to increase or reduce thermal performance of elements, fittings and services, but there are limits – 'the performance of the building fabric and the fixed building services should be no worse than the design limits ...' (paragraph 10 of Part L1A). This is to discourage inappropriate trade-off, e.g. poor insulation offset by renewable energy systems with uncertain lives. However, to demonstrate the required improvement over the energy performance of a notional building, Part L1A warns that in most cases the performance of elements will have to be 'significantly better than those set out' in the document (see the note before paragraph 33).

Notes 29. The Government's *Standard Assessment Procedure for Energy Rating of Dwellings*, Defra, 2005.   30. SBEM publications.

Other factors affecting energy performance and, hence, DER are:

- Limiting solar gains in summer – the dwelling must have appropriate passive control measures to limit the effect of solar gains. (Refer to Appendix 6.7 for further information.) The aim is to reduce or eliminate the need for energy-consuming cooling equipment and to rely on natural ventilation, which will therefore help to improve the DER.
- Quality of construction and commissioning – the quality of construction will be checked throughout the works and on completion to confirm that U-values, thermal bridging, services, controls, lighting and passive solar controls are installed and commissioned as specified in the design DER, and that the building air permeability results are satisfactory in order that an Energy Performance Certificate can be issued. If any of these fall short, the option remains to recalculate the DER on the revised values and demonstrate that DER is still better than TER. Note also the provisional arrangements for air pressure test results (Appendix 6.8).
- Operation and maintenance instructions – 'the necessary provisions for energy-efficient operation'.[31] The incorrect operation and maintenance of boilers, heating, cooling, ventilation and so on will adversely affect thermal performance (Appendix 6.18).

The requirement to meet target carbon emissions applies only to new and not existing dwellings.

6.2.3   Help with achieving TER and improving DER

The factors affecting carbon emission calculations are the thermal performance of the thermal envelope, the control of internal and external heat gains, the energy performance of boilers and cooling equipment, length and insulation of pipes, efficacy of lighting and the air permeability of the building.

Some ways of improving performance are cost-effective, such as improving loft insulation; however, paying for extra insulation ceases to have real cost or energy-saving benefit if air leakage is not also reduced.

Some hints as to what could be cost-effective can be found in Appendix A1 of Part L1B, although please note that this is not applicable to new buildings.

Notes   31. Section 3, Part L1A.

Part L1A also recognises the following contributions towards improving energy efficiency:

A **secondary heating** appliance is assumed to provide some of the space heating. The factor is defined by SAP 2005, depending on the primary and secondary heat source combination. If a secondary appliance is actually fitted, its efficiency will be used to calculate DER. If no appliance is actually fitted, then 10% electric contribution is assumed (paragraph 28 of Part L1A) unless a chimney or flue is provided in which case follow the requirements of paragraphs 2.8.6i and ii.

Low and zero carbon (LZC) energy supply systems can make substantial contributions to achieving the TER,[32] although the capital costs of some of these solutions may be too high to be economic.

A building containing multiple dwellings will achieve compliance if each dwelling has a DER no greater than the corresponding TER or if the average DER is no greater than the average TER (in the latter case, it will still be necessary to provide information for each dwelling).

U-values for elements of the building fabric must be at least as good as those set out in Table 2 of Part L1A.

Design air permeability must be 10m^3/(h·m^2) @ 50Pa or better (see Appendix 6.8).

As part of the final DER calculation, **air permeability** will be based on the results of **pressure tests** (see Appendix 6.8).

Heating and hot water systems must be at least as efficient and meet the minimum control requirements that are recommended in the *Domestic Heating Compliance Guide*, NBS, 2006 (see Appendices 6.9 and 6.10).

Insulation to pipes, ducts and vessels must be provided to standards not less than those in the *Domestic Heating Compliance Guide*, NBS, 2006 (Appendix 6.11).

Mechanical **ventilation** systems must perform no worse that those in GPG 268 (see Appendix 6.12).

Mechanical **cooling** must have an energy efficiency equal to or better than class C in schedule 3 of the labelling scheme under Statutory Instrument 2005 No. 1726 (see Appendix 6.13).

For **fixed internal lighting** the DER should be calculated assuming that 30% of lighting is low energy. Energy-efficient lighting should be installed in the most frequented areas at not less than one per 25m^2 (excluding garages) or one per four fixed fittings (see Appendices 6.14 and 6.15).

Notes

32. *Low and Zero Carbon Energy Sources: Strategic Guide*, DCLG website.

For **fixed external lighting** (supplied from the dwelling supply, i.e. not communal), do not exceed 150W per fitting and provide automatic switching when not required (see Appendix 6.14).

Or compliance for **fixed lighting** whether internal or external would be shown by providing fittings that only take lamps having a luminous efficacy greater than 40 lumens/circuit-watt (see Appendices 6.14 and 6.15).

Fit fluorescent or compact fluorescent light fittings to achieve the above compliances, but not GLS tungsten or tungsten halogen fittings.

Appropriate measures to limit solar heat gains in summer, thus reducing or eliminating the need for air-conditioning and consequently improving the DER (see Appendix 6.7).

Care must be taken with this area of the design process as one measure will often have an impact on other areas, e.g. smaller windows will mean more use of electric lighting (see BS 8206-2[33] for guidance on adequate levels of daylighting).

The TER is based on a notional building with modest amounts of glazing. Buildings that allow greater solar gain will have to compensate through enhanced energy efficiency measures in other aspects of the design (see Appendix 6.7).

Reasonably continuous insulation should be provided over the whole building envelope, avoiding thermal bridges caused by gaps and therefore improving DER. Refer to *Limiting Thermal Bridging and Air Leakage: Robust Construction Details for Dwellings and Similar Buildings, Amendment 1*, TSO, 2002; to BRE Information Paper 1/06 *Assessing the Effects of Thermal Bridging at Junctions and Around Openings*, 2006; and also to BR 262 *Thermal Insulation: Avoiding Risks*, 2001, for advice on how to design details and junctions to reduce the risk of thermal bridging.

Notes   33. BS 8206-2 Code of practice for daylighting.

> The process of achieving the target is likely to be iterative. It would be useful to identify those items that can be modified in the software model fairly easily with little cost or design impact on the building itself, e.g. improve U-values in roofs and walls, try to achieve as much natural ventilation as possible, improve external solar shading design, revise detailing to improve airtightness or provide more efficient heating, cooling or lighting systems.

6.3 Calculating BER (non-domestic)

6.3.1 What is the building emission rate (BER)?

The BER is the weight of carbon dioxide (CO_2) emitted by the actual building per m² of floor area. BER is calculated only for non-domestic buildings, and for dwellings over 450m².

The BER must be calculated with the same NCM approved tool as the TER. The preliminary calculation (optional) is usually carried out as part of the design submission based on plans and specifications and a final calculation demonstrating that the finished building complies with Part L.

Note that the calculation must be carried out with SBEM[34] (see Appendix 6.5) or another modelling tool complying with the requirements of the NCM.

6.3.2 Criteria for demonstrating compliance

Regulation 17C states that 'any new building shall meet the target CO_2 emission rate'.

The five criteria to demonstrate compliance are:

1. achieving an acceptable BER: the predicted rate of carbon dioxide emissions from the building (the BER) is not greater than the target rate (the TER);
2. limits on design flexibility: the TER is unlikely to be achieved unless the performance standards described in Part L2A are exceeded;
3. limiting solar gains in summer: the aim is to reduce or eliminate the need for cooling plant while still maintaining comfortable internal conditions and therefore help to improve the BER;
4. ensuring the quality of construction and commissioning: the performance of the building, as built, is consistent with the prediction made in the BER;

Notes

34. SBEM publications and non-domestic calculation methodology for Part L, DCLG, in preparation or update to CIBSE TM33.

5. providing information: the log book should include data used to calculate the TER and BER.

6.3.3   Help with achieving TER and improving BER

The following means can be used to improve energy efficiency:

Grid-displaced electricity, i.e. generated by photovoltaic, combined heat and power, etc. These emissions will be deducted from total CO_2 emissions before determining BER.[35]

The use of **low and zero carbon energy supply systems (LZC)** can make substantial and economically viable contributions towards meeting the TER.[36]

Management features offer improved energy efficiency. Features such as power factor correction equipment and automatic monitoring with alarms will improve BER by a significant amount.[37]

Appropriate controls should be provided to achieve reasonable energy efficiency as follows:

* building subdivision into control zones of differing solar exposure, or pattern, or use;
* independent timing, temperature control, ventilation and recirculation rates in zones;
* heating and cooling should be controlled to operate independently;
* central plant should operate when the zone system requires, i.e. default is *off* (see Appendix 6.10).

Appropriate lighting design, controls and management, such as avoiding unnecessary lighting of spaces when daylight is sufficient, accessible manual switching, dimmers that reduce rather than divert supply, separate switching of daylit spaces, auto-switching with occupancy or daylight sensors (see Appendices 6.14 and 6.15).

Appropriate energy metering must be provided because 5–10% of the energy being metered can be saved as a result of monitoring energy use.[38]

Relaxing the glazing U-value, which is allowed in some buildings with high internal gains, will reduce the CO_2 emissions and hence the BER (paragraph 38 of Part L2A). However, the case will have to be made to the Building Control body, and in any event the U-value should be no worse than 2.7W/m^{2}K.

Reasonable insulation to pipes, ducts and vessels will reduce heat loss, etc., hence improving BER.[39] **Reasonably continuous insulation** should be provided over the whole building envelope, avoiding thermal bridges caused by gaps, therefore improving BER (see Appendix 6.11).

Notes

35. Note 1, Table 2, p. 6, Part L2A.
36. *Low and Zero Carbon Energy Sources, Strategic Guide*, DCLG website.
37. Table 3 of Part L2A.
38. *Metering Energy Use in Non-domestic Buildings*, GIL 65, Action Energy, 2004.
39. *HVAC Insulation Guide*, TIMSA, in preparation.

Carbon emissions associated with **cooling plant** can be severe so careful attention to management of these systems to reduce or eliminate use is a very effective way of reducing BER.

Appropriate measures to limit solar heat gains in summer, thus reducing or eliminating the need for air-conditioning and consequently improving the BER, include size and orientation of the glazed areas, tints, films and coatings in/on the glass, blinds and shading systems such as overhangs, side fins and brise soleils and using thermal capacity coupled with night ventilation (see Appendix 6.7).

> Care must be taken with this area of the design process as one measure will often have an impact on other areas, e.g. smaller windows will mean more use of electric lighting (see BS 8206-2 for guidance on adequate levels of daylighting).

The TER is based on a notional building with modest amounts of glazing. Buildings that allow greater solar gain will have to compensate through enhanced energy efficiency measures in other aspects of the design.

As part of the final BER calculation, **air permeability/pressure, duct leakage and fan performance** tests will be carried out. Guidance on how to achieve a reasonable design air permeability of 10m³/(h·m²) @ 50Pa is given in *Limiting Thermal Bridging and Air Leakage: Robust Construction Details for Dwellings and Similar Buildings, Amendment 1*, TSO (2002) (see Appendix 6.8).

Ductwork should be made and assembled so as to be reasonably airtight[40] (see Appendix 6.8).

Note that in buildings of area not greater than 500m² air pressure testing can be avoided if the air permeability figure used in the BER calculation is 15m³/(h·m²) @ 50Pa (improvements to the performance of fabric or services will have to be made in order to meet the TER in this case).

The **compliance checklist** is reproduced in Appendix A of Part L2A and is a useful guide when compiling evidence to demonstrate compliance with Part L to building control. This includes the checks, the evidence and where to get it and who can produce it[41] (see Appendix 6.16 of this guide).

Notes

40. *Specifications for Sheet Metal Ductwork*. DW/144, HVCA, 1998.
41. Appendix A in Part L2A.

6.4 What is SAP?

6.4.1 What is the standard assessment procedure (SAP)?

The standard assessment procedure (SAP) is the Government's procedure for rating the energy of dwellings. SAP was designed to be included in the 1995 Building Regulations, and it is now a compulsory component in Part L of the Regulations. A SAP rating is required for all new-build dwellings and those that are created from material change of use. It is optional for extensions.

- The procedure for calculating SAP is defined by the published SAP worksheet which can be downloaded from projects.bre.co.uk/sap2005/. Although the forms can be filled in manually, it is recommended that the calculation should be carried out by a computer program approved by the Building Research Establishment (BRE).
- The SAP provides a simple means of estimating the energy efficiency performance of dwellings. SAP ratings are expressed on a scale of 1 to 100. The higher the number, the better the rating. It can be likened to 'miles per gallon' when comparing fuel consumption in cars.
- The calculation uses the BRE's Domestic Energy Model (BREDEM) to predict heating and hot water costs. The major factors affecting the SAP are the insulation and airtightness of the dwelling and the efficiency and control of the heating system.
- SAP programs are used to enter data on the size of the house, its insulation levels, ventilation system and heating/hot water systems.
- The SAP rating can then be submitted for Building Regulations approval and checking by the local building control department.

6.4.2 Information required by SAP

In order to calculate the SAP, the model needs the following information on the dwelling and environment. This includes:

- materials used for construction of the dwelling, including roofs and walls;
- thermal insulation; this includes the type of thermal insulation materials used in walls, floors or roof construction;
- ventilation characteristics of the dwelling and ventilation equipment; this includes information on extract fans, opening lights, etc.;
- efficiency and control of the heating system(s); boiler efficiency can drastically affect carbon dioxide emissions;
- solar gains through openings in the dwelling; orientation can influence solar gain and this information is needed for a SAP calculation;
- the fuel used to provide space and water heating, ventilation and lighting; fuel costs and environmental impact are taken into account when calculating the SAP rating;
- use of renewable energy technologies.

6.4.3   Output from SAP

The output from the SAP calculation for a dwelling includes:

- Dwelling CO_2 emission rate (DER) (kg CO_2/m²/year): equal to the annual CO_2 emissions per unit floor area for space heating, water heating and ventilation and lighting less the emissions saved by energy generation technologies, expressed in kg/m²/year.
- Energy requirements (kWh/year). The amount of energy required for the heating system.
- Fuel costs. These are calculated using the fuel prices given in the SAP documents. The fuel prices given are averaged over the previous 3 years and across regions.
- SAP rating. The SAP rating consists of a scale from 1 to 100, where bigger is better.

6.4.4   Further information

Further information on the SAP can be found in the following documents:

1. Building Research Establishment. *SAP 2005 Specification*. BRE, 2005.
2. Building Research Establishment. *The Government's Standard Assessment Procedure for Energy Rating of Dwellings*, 2005 edn. BRE, 2005.
3. Building Research Establishment. *Final SAP 2005 Tables*. In: *The Government's Standard Assessment Procedure for Energy Rating of Dwellings, 2005*. BRE, 2005.
4. Building Research Establishment. *Final SAP 2005 Worksheet*, 2005 edn. BRE, 2005.

6.5   Evaluating energy use and carbon emissions in buildings through the Simplified Building Energy Model (SBEM)

6.5.1   What is the Simplified Building Energy Model?

The Simplified Building Energy Model (SBEM) is run on an MS Access database and is intended to provide evaluations of energy consumption and carbon emissions in non-domestic buildings as well as domestic buildings over 450m².

It was developed by BRE as part of the NCM for the then Office of the Deputy Prime Minister (ODPM), now the Department for Communities and Local Government (DCLG).[42]

The reports produced by SBEM assist with the design process and eventually demonstrate that mandatory carbon emission targets have been met under Part L for building performance certification purposes.[43]

Notes

42. www.ncm.bre.co.uk
43. Section 2.1: A User Guide to iSBEM, ODPM/ BRE, 24 January 2006.

The software can be downloaded from http://www.ncm.bre.co.uk/
download.jsp together with a tutorial manual, although MS Access 2000
or a later version is required to run it.

> The NCM allows the actual calculation to be carried out either
> by SBEM or by other approved modelling software. SBEM is
> not an appropriate method of calculation in all cases.

6.5.2   Before using SBEM

The inputting screens are not particularly intuitive and take a little getting
used to. It is sensible to 'play' with the software before using it in earnest
because it is, for example, all too easy to overwrite your records. Playing
with the software will also help you to find appropriate 'notional' values
buried in the picklists (in the 'project' form) for calculating the TER and
to find the gaps where it may be necessary to use default data from the
tables in Part L or to carry out some research.

First, use Appendix A of the iSBEM user guide to identify the kind of
information required in order to use the SBEM software. There are useful
tips on naming conventions in Chapter 8 of the guide.

Then before entering the data to calculate the TER for a notional building,
it is necessary to identify from drawings or survey the various activity
zones or zones that have different thermal gains, independent heating
or cooling services. Each of these zones will have a defined geometry and
orientation which must be set up in the system so that the quantity and
orientation of elements, fittings and services can be associated with these
zones.

6.5.3   Summary of the SBEM evaluation process

- Gather the required data for the building.
- Analyse the information and identify the different zones.
- Enter the information into the model and run the calculation.[44]

6.5.4   Information required by SBEM

In order to calculate the energy use and carbon emissions using SBEM, it is
necessary to provide information on the following topics.[45]

General information
The first step is to provide general project information including weather
(defaults in SBEM are London, Manchester and Edinburgh), building use,
ownership and certifier details.

Notes

44. Section 8.1: A User Guide to iSBEM, ODPM/
    BRE, 24 January 2006.
45. Section 8: A User Guide to iSBEM, ODPM/
    BRE, 24 January 2006.

Construction information

The model needs information about the different types of construction used in the building. All of the material information about all types of walls, floors, roofs, doors and glazing in the building is required. This information is then available for selection from a picklist when defining the building geometry. When calculating the TER, use the 'notional' construction for the appropriate version of Part L selected from the picklist. (Later when calculating BER, it is necessary to select the actual materials, which are likely to have to perform in a way that is equal to or better than the notional version.)

Geometry and zoning information

After filling in a field highlighted in green, pressing the 'enter' button allows editing of the entries in the other boxes on the form.

The model contains data on U-values for most standard forms of construction and uses this along with geometry and zoning information to calculate energy losses.

An activity area is defined as an area that is assigned its own comfort conditions, standard operating pattern and heat gains profile.

Zones need to be defined from analysis of drawings, schedules or survey.[46] The relationships between zones will also have to be considered along with air permeability, thermal bridging, glazing types and any shading systems used.

Building services information

The model also needs information about all the building systems, including heating and ventilation, hot water, sustainable energy technologies and combined heat and power. Further information on the lighting and ventilation characteristics for each zone is also required, for example:

- heating, ventilation and air-conditioning, including quantity, fuel, efficiency, heat source, cooling system and ventilation, metering, control and adjustments;
- domestic hot water, including quantity, fuel, generator type, storage and secondary circulation;

Notes

46. Sections 8.3 and 8.4: A User Guide to iSBEM. ODPM/BRE, 24 January 2006.

- sustainable energy, including quantity, type, size, inclination and orientation for a solar-powered or photovoltaic system and quantity, terrain, diameter, hub height and power for a wind-powered system;
- combined heat and power, including fuel, efficiency and various ratios between space heating and water and so on.

The various building systems must then be assigned to their zones. Information required includes pipe lengths, ventilation type, heat recovery, fan power and lighting design, efficiency, lamp type, switching, controls and sensors.

6.5.5   Ratings and outputs generated by SBEM[47]

Part L ratings and compliance checking
- Notional building emission rate: $kgCO_2/m^2$ for the notional building.
- Target emission rate (TER): ($kgCO_2/m^2$) this is the notional building emission rate with two factors applied to it – improvement factor (IF) and low or zero carbon factor (LZC).
- Building emission rate (BER): $kgCO_2/m^2$ for the actual building.
- Pass or fail CO_2 emissions: if the BER < TER, the building passes the CO_2 emissions element of Part L.

Other Part L checks (such as those for U-values) can be found in the documents generated by SBEM.

Asset ratings (and recommendations)
The following information can be provided by SBEM reports:

- The energy used per m^2 (kWh/m^2) by the actual building and the notional building for heating, cooling, auxiliary energy uses (pumps and fans), lighting and domestic hot water.
- The energy used per m^2 (kWh/m^2) by the actual building and the notional building in terms of electricity and fuel use.
- The resulting CO_2 emissions ($kgCO_2/m^2$) from the actual and notional building.
- The percentage of the energy consumed by the notional building that is consumed by the actual building.
- The asset rating (currently the percentage CO_2 emitted by the actual building out of the total virtually emitted by the notional building).

6.5.6   Reports

The **Building Regulations compliance document** will form part of the submission by designers to building control to demonstrate compliance with L2A. The document follows exactly the criteria found in Appendix A Compliance Checklist of L2A, and SBEM completes the appropriate

---

Notes

47. Section 7: A User Guide to iSBEM. ODPM/ BRE, 24 January 2006.

sections of the document. Where an additional submission is required, the document will state that clearly.

The **main output report** gives a summary of the energy performance of the building including:

- details on the building, certifier and building owner;
- whole-building CO_2 performance;
- annual energy consumption by end use;
- monthly energy consumption by end use.

Data reflection reports contain all the data entered into SBEM for the building and are to be attached to the building log book.

Technical output report produces data for use in more in-depth analysis of the results including:

- monthly and annual energy use by fuel type;
- monthly and annual energy use by end use;
- monthly and annual CO_2 emissions;
- energy production and CO_2 displaced by renewables.

6.6   U-values

6.6.1   What is a U-value?

The U-value (thermal transmittance) of a building element is a measure of the rate of heat flow through the element. Technically, it is calculated as the rate at which heat transfers through 1m² of a structure with a temperature difference between the internal and external environments of 1°C. The greater the level of insulation of the building element, the lower its U-value, measured in watts per square metre of area per degree temperature difference across the building element (W/m² K). Lower U-values indicate better thermal insulation. For example, a wall with a U-value of 0.3 W/m²K loses heat at half the rate of a wall with a U-value of 0.6 W/m²K.

6.6.2   Calculation for U-values

Guidance on the use of calculation methods is contained in BR 443 *Conventions for U-value Calculations*.[48]

There are two ways that U-values can be assessed. Either the individual U-value of an element in a particular plane can be calculated or the average U-value for all elements of the same type can be calculated (the area-weighted average U-value). The area-weighted U-value is used for

---

Notes

48. Anderson, B. and Building Research Establishment, *Conventions for U-value Calculations*, BRE Report BR 443, 2006 edn. Construction Research Communications Limited, 2006.

calculating the heat loss from the building; however, there is a limit on the heat loss allowed for an individual element (the limiting U-value).

There are limits on flexibility for several reasons. There is a risk of creating thermal bridges with the consequent risk of condensation forming internally or within vulnerable parts of the construction. Also, it is not sensible to compensate for poor thermal performance of elements of construction by using highly efficient heating systems of uncertain life.

6.6.3   What is a thermal bridge?

The *Moisture Control Handbook*[49] defines a thermal bridge as 'regions of relatively high heat flow conductance in a building envelope'. An example of a thermal bridge is an uninsulated window lintel or the edge of a concrete floor slab built into a solid blockwork wall. Thermal bridges can have a major effect on the thermal performance of building envelopes, significantly increasing winter heat loss and summer heat gain. Further, condensation on thermal bridging elements can result in mould and mildew growth (with accompanying reduction of air quality), staining of surfaces and serious damage to building components.

6.6.4   Assessing cold bridge/thermal bridges

For Building Regulations purposes, one way of demonstrating that provision has been made to limit thermal bridging at junctions and around openings is to use the details in Accredited Construction Details when these are published.

6.6.5   What is condensation?

There are two types of condensation to be concerned about – interstitial condensation and surface condensation. In each case condensation occurs on cold surfaces when warm, moist air comes into contact with them. This happens when the temperature of the surface is below the so called 'dewpoint' temperature of the warm air. As the air in contact with the cold surface is cooled to below its dewpoint, it must release excess moisture that it can no longer support. It releases this moisture as liquid water, which appears on the colder surface.

Interstitial condensation
Most building materials, except metals, plastics and similar materials, are to some extent permeable and do not obstruct the movement of moist air through the structure. The warm moist air can cool below its dewpoint within the fabric of the building, resulting in condensation at a layer or air space. Because the condensation is hidden, it can go undetected for long periods, sometimes resulting in serious damage such as timber decay.[50]

Notes

49. Lstiburek, J. and J. Carmody, *Moisture Control Handbook: Principles and Practices for Residential and Small Commercial Buildings*, London: International Thomson, 1993, xiv, 214 pp.

50. British Standards Institute, *Hygrothermal Performance of Building Components and Building Elements. Internal Surface Temperature to Avoid Critical Surface Humidity and Interstitial Condensation: Calculation methods (AMD Corrigendum 13792)*, British Standards Institute, 2002.

Surface condensation

Surface condensation is familiar to most people. It can lead to mould growth, which can be unsightly and can damage decorative finishes. Mould growth can cause a musty odour and can also be a health hazard.

Assessing interstitial condensation risk

To assess the risk of interstitial condensation we need to know two things – the temperature profile through the structure and the vapour resistance of the various layers

The procedure for assessing the risk of interstitial and surface condensation is described in the British Standard *Code of Practice for Control of Condensation in Buildings.*[51]

Notes

51. BS 5250:2002 *Code of Practice for Control of Condensation in Buildings.*

Summary of all U-values referred to in Part L

	L1A New dwelling			L1B existing dwelling								
	Table 2(a) AW dwelling average	Table 2(b) worst individual subelement	SAP reference value for TER	Table 1(a) AW U-value standards	Table 1(b) limiting U-value standards	Table 2(a) standards for new fittings in extns	Table 2(b) standard for replacement fittings in extn	Table 4(a) standards for TEs	Table 4(b) standards for replacement elements	Table 5(a) threshold value for upgrading retained TEs	Table 5(b) improved value for upgrading retained TEs	Table A1 cost-effective U-values
Walls (opaque panels, curtain walls)	0.35	0.7	0.35	0.35	0.7			0.3	0.35	0.7	0.35	0.35
Cavity wall										0.7	0.55	0.55
Floor	0.25	0.7	0.25	0.35	0.7			0.22	0.25	0.7	0.25	
Roof	0.25	0.35	0.16	0.25	0.35							0.25
Pitched roof – insulation @ ceiling								0.16	0.16	0.35	0.16	0.16/0.20
Pitched roof – insulation @ rafters								0.2	0.2	0.35	0.2	0.16/0.20
Flat roof								0.2	0.25	0.35	0.25	
Window (glazed areas, curtain walls)	2.2	3.3	2.2*	2.2	3.3	1.8†	2.0‡					
Roof window	2.2	3.3		2.2	3.3							
Dormer window side walls												0.35
Rooflight	2.2	3.3		2.2	3.3							
Door	2.2	3.3	2	2.2	3.3	3	3					
Glazed door			2.0*			2.2§	2.2§					
Vehicle access												
High use entrances												
Roof ventilation												
Curtain wall												
Display wndows												

AW, area-weighted; extn, extension; TE, thermal elements

*Double glazed, low-E hard coat, frame factor 0.7, solar energy transmittance 0.72, light transmittance 0.80

†OR Window energy rating Band D (in dwelling or building of domestic character) OR centre pane U-value 1.2 W/m^{2}K

‡OR Window energy rating Band E (in dwelling or building of domestic character) OR centre pane U-value 1.2 W/m^{2}K

§OR centre pane U-value 1.2 w/m^{2}K

| L2A New buildings other than dwellings | | L2B Existing buildings other than dwellings | | | | | | | |
Table 4(a) limiting U-value standards	Table 4(b) for any individual subelement	Table 3(a) AW U-value standards	Table 3(b) limiting U-value standards	Table 5(a) standards for new fittings in extns	Table 5(b) standards for replacement fittings	Table 6(a) standards for new fittings in extn	Table 6(b) standards for replacement fittings in existing building	Table 7(a) upgrade retained TEs, threshold	Table 7(b) improved value for upgrading TEs
0.35	0.7	0.35	0.7			0.3	0.35**	0.7	0.35
								0.7	0.35
0.25	0.7	0.25	0.7			0.22	0.25§	0.7	0.25
0.25	0.35	0.25	0.35						
						0.16	0.16	0.35	0.16
						0.2	0.2	0.35	0.2
						0.2	0.25	0.35	0.25
2.2	3.3	2.2	3.3	1.8†	2.2‡				
2.2	3.3	2.2	3.3	1.8†	2.2‡				
2.2	3.3	2.2	3.3	1.8†	2.2‡				
2.2	3	2.2	3.3						
				2.2	2.2§				
1.5	4			1.5	1.5				
6	6			6	6				
6	6			6	6				
No limit				Exempt	Exempt				

6.6.7 Sample U-values for common compliant forms of construction
Roofs

Figure 6
Roof 1: Cold roof construction

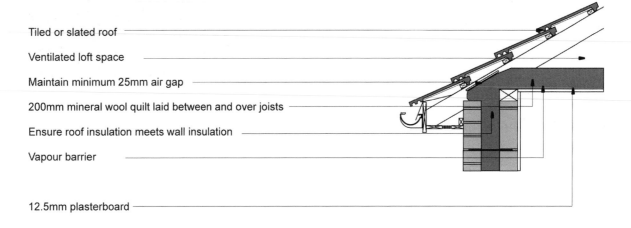

Tiled or slated roof

Ventilated loft space

Maintain minimum 25mm air gap

200mm mineral wool quilt laid between and over joists

Ensure roof insulation meets wall insulation

Vapour barrier

12.5mm plasterboard

U-value = 0.13Wm²K

Thickness (mm)	Major material	Conduct. (W/mK)	Minor material	Conduct. (W/mK)	Resistance (m²K/W)	Vapour resistance (MNs/g)
25	Concrete (2000 kg/m²)	1.35			0.02	1.5
25	96% cavity (ventilated)	0.09	4% timber – softwood	0.13	0.09	0.15
1	Bitumen – felt/ sheet	0.23			0	50
1000	Cavity (ventilated)	0.09			0.09	5.3
170	Mineral wool quilt	0.042			4.05	0.85
150	93% mineral wool quilt	0.042	7% timber – softwood	0.13	3.11	0.91
1	Polyethylene/ polythene	high–0.500			0	100
13	Plasterboard	0.25			0.05	0.39
n/a	Internal resistance	n/a			0.1	n/a

The information in this table was produced using NHER Evaluator version 4.10 (2003) for illustrative purposes only.

Figure 7
Roof 2: Transverse section across pitched roof (room in roof)

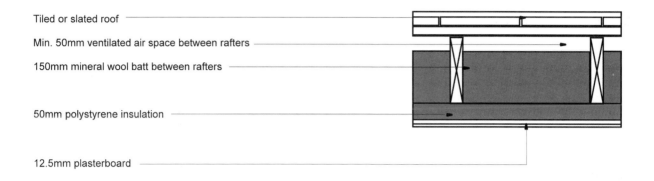

Tiled or slated roof

Min. 50mm ventilated air space between rafters

150mm mineral wool batt between rafters

50mm polystyrene insulation

12.5mm plasterboard

U-value = 0.15Wm²K

Thickness (mm)	Major material	Conduct. (W/mK)	Minor material	Conduct. (W/mK)	Resistance (m²K/W)	Vapour resistance (MNs/g)
n/a	External resistance	n/a			0.04	n/a
25	Concrete (2000 kg/m²)	1.35			0.02	1.5
25	96% cavity (ventilated)	0.09	4% timber – softwood	0.13	0.09	0.15
1	Breather membrane	0.17			0.01	2000
38	Cavity (unventilated)	0.18			0.18	0.2
80	Mineral wool batt	0.038			2.11	0.4
100	93% mineral wool quilt	0.042	7% timber – softwood	0.13	2.08	0.6
50	Polyurethane board	0.025			2	5.75
1	Polyethylene/ polythene	0.5			0	100
13	Plasterboard	0.25			0.05	0.39
n/a	Internal resistance	n/a			0.1	n/a

The information in this table was produced using NHER Evaluator version 4.10 (2003) for illustrative purposes only.

Figure 8
Roof 3: Flat roof with non-ventilated void (warm roof)

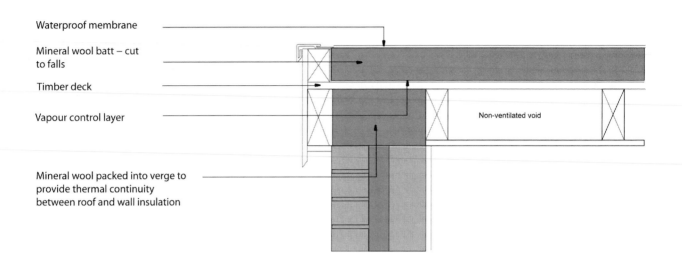

Waterproof membrane

Mineral wool batt – cut to falls

Timber deck

Vapour control layer

Mineral wool packed into verge to provide thermal continuity between roof and wall insulation

Non-ventilated void

U-value = 0.25Wm²K

Thickness (mm)	Major material	Conduct. (W/mK)	Minor material	Conduct. (W/mK)	Resistance (m²K/W)	Vapour resistance (MNs/g)
n/a	External resistance	n/a			0.04	n/a
2	Bitumen – felt/ sheet	0.23			0.01	100
130	Mineral wool batt	0.038			3.42	0.65
1	Polythene	0.5			0	99,999
18	Oriented strand board (OSB)	0.13			0.14	0.54
200	90% cavity (unventilated)	0.18	10% timber – softwood	0.13	0.2	1.35
13	Plasterboard	0.25			0.05	0.39
n/a	Internal resistance	n/a			0.1	n/a

The information in this table was produced using NHER Evaluator version 4.10 (2003) for illustrative purposes only.

Walls

Figure 9
Block 1: Rendered blockwork with fully filled cavity

19mm render
100mm AAC blockwork
75mm cavity with full-fill mineral wool batt insulation
100mm AAC blockwork
15mm plaster
Cavity tie

Structural temperature ————
Dewpoint temperature — — —

U-value = 0.32Wm²K

Thickness (mm)	Major material	Conduct. (W/mK)	Minor material	Conduct. (W/mK)	Resistance (m²K/W)	Vapour resistance (MNs/g)
n/a	External resistance	n/a			0.06	n/a
19	Rendering (external)	0.57			0.03	1.33
100	93% AAC block	0.18	7% mortar (exposed)	0.94	0.43	1.71
75	Mineral wool batt	0.038			1.97	0.38
100	93% AAC block	0.18	7% mortar (protected)	0.88	0.44	1.71
15	Gypsum (1500 kg/m²)	0.56			0.03	0.06
n/a	Internal resistance	n/a			0.12	n/a

The information in this table was produced using NHER Evaluator version 4.10 (2003) for illustrative purposes only.

Figure 10
Brick 1: Brick faced cavity wall with insulated dry-lining

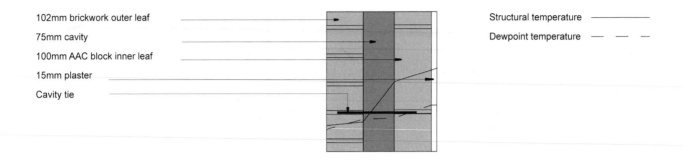

102mm brickwork outer leaf
75mm cavity
100mm AAC block inner leaf
15mm plaster
Cavity tie

Structural temperature ——————
Dewpoint temperature — — —

U-value = 0.29Wm²K

Thickness (mm)	Major material	Conduct. (W/mK)	Minor material	Conduct. (W/mK)	Resistance (m²K/W)	Vapour resistance (MNs/g)
n/a	External resistance	n/a			0.06	n/a
100	87% brickwork (outer leaf)	0.77	13% mortar (exposed)	0.94	0.13	4.5
100	Mineral wool batt	0.038			2.63	0.5
100	93% AAC block	0.18	7% mortar (protected)	0.88	0.44	1.71
15	Gypsum (1500 kg/m²)	0.56			0.03	0.06
n/a	Internal resistance	n/a			0.12	n/a

The information in this table was produced using NHER Evaluator version 4.10 (2003) for illustrative purposes only.

Figure 11
Brick 2: Brick faced cavity wall with insulated dry-lining

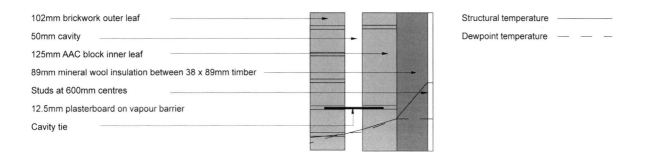

102mm brickwork outer leaf

50mm cavity

125mm AAC block inner leaf

89mm mineral wool insulation between 38 x 89mm timber

Studs at 600mm centres

12.5mm plasterboard on vapour barrier

Cavity tie

Structural temperature

Dewpoint temperature

U-value = 0.33Wm²K

Thickness (mm)	Major material	Conduct. (W/mK)	Minor material	Conduct. (W/mK)	Resistance (m²K/W)	Vapour resistance (MNs/g)
n/a	External resistance	n/a			0.06	n/a
100	87% brickwork (outer leaf)	0.77	13% mortar (exposed)	0.94	0.13	4.5
50	Cavity (unventilated)	0.18			0.18	0.27
125	93% AAC block	0.18	7% mortar (protected)	0.88	0.55	2.14
89	92% mineral wool batt	0.038	8% timber – softwood	0.13	1.96	0.55
1	Polythene	0.5			0	99,999
15	Plasterboard	0.25			0.06	0.45
n/a	Internal resistance	n/a			0.12	n/a

The information in this table was produced using NHER Evaluator version 4.10 (2003) for illustrative purposes only.

Figure 12
Brick 3: Brick faced and 50mm cavity and timber frame

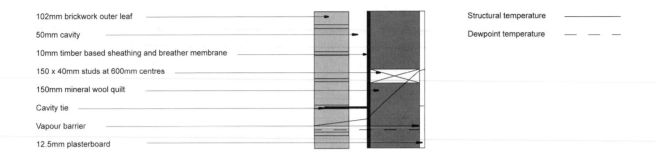

102mm brickwork outer leaf
50mm cavity
10mm timber based sheathing and breather membrane
150 x 40mm studs at 600mm centres
150mm mineral wool quilt
Cavity tie
Vapour barrier
12.5mm plasterboard

Structural temperature
Dewpoint temperature

U-value = 0.3Wm²K

Thickness (mm)	Major material	Conduct. (W/mK)	Minor material	Conduct. (W/mK)	Resistance (m²K/W)	Vapour resistance (MNs/g)
n/a	External resistance	n/a			0.06	n/a
100	87% brickwork (outer leaf)	0.77	13% mortar (exposed)	0.94	0.13	4.5
50	Cavity (unventilated)	0.18			0.18	0.27
1	Breather membrane	0.17			0.01	2000
10	Fibreboard (600kg/m²)	0.14			0.07	0.12
150	85% mineral wool quilt	0.042	15% timber – softwood	0.13	2.72	1.09
1	Polythene	0.5			0	99,999
15	Plasterboard	0.25			0.06	0.45
n/a	Internal resistance	n/a			0.12	n/a

The information in this table was produced using NHER Evaluator version 4.10 (2003) for illustrative purposes only.

6.7 Checking for solar gains and overheating

6.7.1 Summer overheating

As insulation levels reduce heat demand, paradoxically the risk of overheating increases. There are several causes for it. Although solar gain is the primary cause of external heat gain, many causes are internal, e.g. from cooking and hot water systems, appliances, lighting, computers and people.

The introduction of cooling systems increases energy consumption and a balance of measures will be needed to achieve the TER.

In dwellings, the requirement is that provision should be made to prevent high internal temperature due to solar gain (paragraph 46 of Part L1A).

There is no explicit requirement for dealing with summer overheating when carrying out work to an existing dwelling under Part L1B; however, this will be a consideration if using SAP calculations to demonstrate the improved energy performance of the building in connection with providing an extension or works to create a new dwelling through a **material change of use.**

In buildings other than dwellings, Part L (paragraph 60 of Part L2A) requires that provision should be made to limit solar gain in occupied spaces that do not have comfort cooling (unless the space is a stack or unoccupied atrium used to drive air circulation).

In existing buildings other than dwellings, it may be necessary to consider measures to control solar gain in order to comply with the provisions for a **consequential improvement** resulting from the introduction of a new cooling installation (or if the cooling capacity per unit area of an existing system is being increased). Paragraph 25(b) of Part L2B sets out the requirements to improve the solar gain co-efficient if there is greater than 40% existing window area and greater than 20% rooflight area. The additional strategy in Part L2B is to reduce the internal heat gain from lighting (paragraph 25(c) of Part L2B).

Apart from this there is no explicit requirement for dealing with summer overheating when carrying out work to an existing building under Part L2B; however, solar gain will be a consideration if using energy modelling calculations to demonstrate the improved energy performance of the building in connection with providing an extension or works that are a **material change of use.**

The provisions of Part F will also have to be considered in terms of the contribution to purge and background ventilation of windows.

6.7.2 Reducing the impact of overheating through solar gain

Solar gain is reduced by a combination of choosing an appropriate window size, orientation and shading, and by using high thermal capacity coupled with night ventilation. (Night air is ventilated into the building to cool, for example, the soffit of an internal concrete floor slab that has high thermal capacity so that during the day the ventilation is closed and the cooled floor slab absorbs heat from the space.) External shading (e.g. blinds, awnings, brise soleils, shutters, louvres, balconies over lower storeys or managed vegetation) is much more effective than internal shading at reducing the impact of solar gain, especially when low-E glass is specified. Useful advice can be found in *Solar Shading of Buildings BR 364*, CRC Ltd, 1999.

> Managed vegetation cannot be relied upon as an effective method of solar shading and may also reduce the amount of daylight below acceptable levels.

Careful orientation of windows at design stage can give benefits in both summer and winter. South-facing glazing can be shaded from high summer sun, but winter sun can pass through the glass and penetrate into the space, increasing solar gain in the winter months and offsetting heating bills. Even without shading there is a benefit because of the differing angles of incidence of the sun's rays on the glass. Large amounts of west-facing glazing should be avoided.[52]

> The shading will have a negative influence on available daylight levels.
>
> Window size not only affects solar gain but it also has an impact on lighting requirements. A small window may reduce thermal gain from outside, but at the same time the increase in lighting requirement will increase the heat gain from internal lighting systems as well as increase energy consumption. BS 8206 Part 2 Code of practice for daylighting and *Designing with Rooflights to Satisfy ADL2 (2006)*, NARM Technical Guidance (2006) are referred to in Part L2A for guidance.

Other factors that influence solar gain are as follows:

- buildings facing due south perform better than those facing south west;[53]

Notes

52. *Energy Efficiency Best Practice in Housing: Reducing Overheating – a Designer's Guide*, The Energy Saving Trust.

53. *Energy Efficiency Best Practice in Housing: Reducing Overheating – a Designer's Guide*, The Energy Saving Trust.

- increasing the thermal mass and increasing the amount that is exposed to the internal environment reduces the effect of overheating; this is more difficult with lightweight construction such as timber frame or in heavy construction where insulation is applied internally;
- increasing night ventilation and decreasing day ventilation helps keep gains down; the benefit of doing this must be included in instructions to the occupier.

6.7.3 Reducing the impact of overheating through internal gain

Increase the efficacy of lighting systems (see Appendix 6.24).

Insulate hot water systems and pipework and reduce pipe runs to the minimum possible.

Insulate cooling plant and ductwork to prevent heat gain from the internal environment itself.

Encouragement is given to go beyond the requirements of the Building Regulations and reference is made to *Climate Change and the Indoor Environment: Impacts and Adaption,* CIBSE, TM36.

A useful source of advice for domestic situations is the Energy Savings Trust publication *Reducing Overheating: a Designer's Guide,* CE129.

Paragraph 64 of Part L2A describes target levels of comfort for spaces that are not air-conditioned. Basically, internal and external heat gains should not exceed 35W/m^2 and the internal dry bulb temperature should not exceed 28°C for more than a reasonable number of hours (20 hours is suggested), although there are more specific requirements for schools (paragraph 62 of Part L2A).[54]

As pointed out in the subnote to paragraph 65 of Part L2A, the TER is based on a notional building with modest amounts of glazing. Increased glazing can be compensated for by the provision of comfort cooling systems, more sophisticated solar control measures and better thermal efficiency measures elsewhere; however, a balance must be struck to keep the actual building performance below the TER.

Paragraph 61(a) L2A makes reference to a table of 'design irradiances' in *Environmental Design*, CIBSE Guide A. These tables are aimed at experts and their explanation and use is probably best left to the M&E consultant. (Generally the tables, set out by latitude in five-degree increments, show a range of 'clear day beam and diffuse irradiances' on vertical and horizontal surfaces in specified days in the northern (N) and southern (S) hemispheres in W/m^2 and include a daily 'mean'. Table 2.24 referred to in

---

Notes

54. *Guidelines for Environmental Design in Schools*, Building Bulletin 87 and Building Bulletin 101 (regularly updated to meet the latest requirements of Part L and Part F).

Part L2A provides this data for London (Bracknell). The combined internal and external gains using this table should not exceed 35W/m².)

6.8   Designing buildings to control air permeability

6.8.1 What is air permeability?

Air permeability is the uncontrolled leakage of air through the thermal envelope of the building. This uncontrolled and unplanned air flow reduces the effectiveness of the insulated elements and accounts for half of the heat losses from building fabric in the UK[55] and adds to the burden on heating and cooling systems (adding up to 40% in energy costs).[56] Controlled air ventilation is necessary and the requirements are covered in Part F (refer also to *Guide to Part F*. NBS, 2006).

Part L requires that this uncontrolled air leakage is reduced to an acceptable level and that the results of air permeability tests are taken into account when calculating the energy performance of a building.

Part L documents refer for guidance to *Limiting Thermal Bridging and Air Leakage: Robust Construction Details for Dwellings and Similar Buildings*, Amendment 1, TSO (2002). Further guidance can be obtained from SEDA *Design and Detailing for Airtightness*, which is published on the Internet at www.seda2.org. Guidance for cladding systems is also given in the MCRMA/EPIC technical report.[57]

The air leakage test is described in CIBSE *Technical Memorandum TM23:2000* (see Appendix 6.8 of this guide). However, the designer needs to ensure that uncontrolled air leakage from the structure will be less than assumed for the energy performance calculations in order to pass the air leakage test (subject to the correct quality of workmanship on site being achieved) and so meet the criterion for issue of the Energy Performance Certificate. Alternatively, a certificate can be issued as long as an air leakage is less than 10m³/(h·m²) and BER is less than TER.

6.8.2   Causes of air leakage

Air leakage occurs through porous wall constructions, especially mortar joints, at the edges of windows, doors, panels, cladding and around structures or services that penetrate the external envelope of the building.[58] Other routes for air leakage include gaps in curtain walling, at eaves and so on, as well as gaps in windows, under doors, around loft hatches and flush light fittings, poorly fitted ducted extracts, between floorboards and so on. There are obvious routes for air leakage such as chimney flues.

Notes

55. BRE, *Airtightness in Commercial and Public Buildings*, 2002.
56. BRE, *Airtightness in Commercial and Public Buildings*, 2002.
57. *Guidance for Design of Metal Roofing and Cladding Systems to Comply with Approved Document L2 (2006)*, MRMCA Technical Paper No. 17 and EPIC, 2006.
58. *CIBSE Technical Memoranda*, TM23

6.8.3   Design to reduce air leakage

The ultimate responsibility for meeting the requirements of air permeability lies with the constructor. However, depending on the form of procurement, the designer will need to provide carefully considered details and performance specifications. Some of the design considerations are as follows.

Consider the appropriate form of construction for reducing air permeability in the early stages of design.

Define the heated/cooled zones and, for air leakage purposes, treat them as separate from unheated zones such as service corridors, vertical shafts, boiler rooms and plant rooms.

Dry forms of construction require particular attention to seal joints between panels, allowing for shrinkage and movement between rigid elements. Flexible joints and seals deteriorate over time and access should be provided for inspection and replacement to maintain performance. Curtain walling design needs particular attention to the joints between panels and the frame and also where the curtain walling meets other elements such as the roof and abutting walls. In these cases, a combination of membrane lapped with vertical and horizontal damp-proof membranes together with a mastic seal may be necessary.

Blockwork is porous and walls that separate zones with different energy performance requirements should be rendered or plastered. Membranes and damp-proof courses should be lapped and sealed.

All junctions between hard materials (e.g. skirting to floor, wall to plasterboard ceiling) will probably need to be sealed with mastic or gaskets that can be easily replaced. Cavity closing details around windows (including the cill detail) and other thermal envelope penetrations should be carefully sealed to avoid air infiltration. Expanding foam gap filler could be used.

Seal around holes through concrete floor slabs. Use joist hangers instead of building joist ends into walls.

Loft hatches are a significant cause of air leakage.[59]

Eaves details are a common path for air leakage and the introduction of a vapour barrier seal here is recommended, taking care not to obstruct the ventilation path to the uninsulated parts of the roof space.

Gaps in insulation panels in walls are not a cause of air leakage through the thermal envelope but air flow around the insulation can reduce its effective performance substantially and it would be sensible to specify insulation to have taped tongue and groove joints.

Notes

59. BRE Information Paper 01/00, January 2000, found that 2% of air leakage was through loft hatches.

6.8.4   How air leakage testing is undertaken

The required pressure tests on buildings are to BS EN 13829:2001 Thermal performance of buildings. Determination of air permeability of buildings. Fan pressurisation method. Refer to *Testing Buildings for Air Leakage*, CIBSE Technical Memoranda, TM23 (2000).

Test results are plotted as flow volume versus pressure difference. The test should be carried out by a suitably qualified person (usually a member of the Air Tightness Testing and Measurement Association) who confirms in a written report that the approved procedure has been followed and records design and measured air permeability. Air permeability should not be less than $10m^3/(h \cdot m^2)$ at 50Pa, and BER using test results should be better than TER.

Air leakage tests are required on virtually all new buildings in order to demonstrate that the designed values have been achieved in the construction phase. The exceptions to this rule are:

- larger housing developments, where sample dwellings can be tested;
- small housing developments (one or two dwellings only), where reference can be made to either:
  - dwellings of a similar type constructed by the same builder and tested in the previous 12 months, or
  - by using a value for air permeability of $15m^3/h \cdot m^2$ at 50Pa when calculating the DER;
- non-domestic buildings of less than $500m^2$.

If the building is correctly designed (see Appendix 6.8.3), leakage occurs because of failure either to construct the design details correctly or failure to use the specified detail.

Testing to BS EN 13829 is carried out on the completed building with all 'designed' air paths such as windows, ventilators, chimneys and flues blocked off. Cupboard doors can remain closed; all other internal doors should be wedged open. Drainage traps should contain water. Fire dampers and louvres should be closed, mechanical ventilation systems turned off and the inlet and extract grilles blocked unless the system is being used to provide the air pressure. Lift and other plant rooms are normally considered to be outside the tested envelope.

The test methodology is based on the premise that the pressure difference is uniform over the entire building envelope.

For domestic buildings, it is normal for the purpose of the test to replace one of the external doors with a blocking piece in which is mounted a fan unit capable of producing a pressure difference of 50Pa at a flow rate of $1.0m^3/s$. In order to achieve the same pressure in non-domestic buildings, flow rates of up to $30m^3/s$ will be required and a number of high-volume fans may be used. If the building is mechanically ventilated, it may be possible to use its own HVAC system to pressurise the building. If this is the case, the outlet vents should be blocked while the inlet vents remain open. Fan capacity should be determined such that a minimum air volume flow rate of 70% of the design rate at 50Pa pressure difference can be maintained.

The test is carried out in stages with readings being taken at approximately 5Pa steps (minimum five readings) starting at 10Pa and proceeding up to 55–60Pa. Further readings are taken as the pressure is progressively reduced. If the building is so leaky that the full pressure difference cannot be achieved, extrapolated results can be used provided that a difference of 25Pa can be achieved. Below this level, the results will not be valid.

The effectiveness of any temporary seals should be checked during the course of the test. Failures are usually evident as a result of anomalous readings.

6.8.5 Interpreting the results of an air permeability test

If a test result shows that leakage is higher than anticipated, it is recommended that suspect details are selectively blanked off temporarily and the test re-run. This should give valuable information with regard to the design of remedial measures.

Test results are plotted as flow volume versus pressure difference; the requirement is that both the upward and downward curves appear to be reasonably smooth. If this is the case, the 50Pa value (or extrapolated value) represents the test result and should be divided by the area (m^2) of the building envelope being tested (including the ground floor) to give the air permeability in m^3/h·m^2 of building envelope.

If a building fails to achieve its design air permeability, it may still meet the requirements if it achieves both the target air permeability and dwelling (or building for buildings other than dwellings) emission rate. A construction failing on both of these criteria must be retested following the implementation of remedial measures.

A relaxation of these rules applies until 31 October 2007, whereby a retested building must show a minimum improvement of the lesser of

EITHER 75% of the identified shortfall between design and test values

OR a retest result within 15% of the design requirement

AND the DER or BER is better than the amended TER to include the measured permeability.

6.9 What is an efficient boiler?

Boiler efficiency is not a constant and varies according to the season of the year and the varying loads placed on the boiler. The standard system of tests appropriate to the UK is known as the SEDBUK (seasonal efficiency of domestic boilers in the UK) method. All new gas and oil boilers are rated by an independently certified test and placed on the boiler efficiency database at www.sedbuk.com. The list is updated monthly.

The efficiency of new boilers varies between about 70% and 90%. Condensing boilers are more efficient and generally exceed 80% efficiency but may be more expensive to purchase. The *Domestic Heating Compliance Guide*, NBS (2006), states for gas-fired boilers 'that the boiler efficiency [in new buildings] should not be less than 86%'.

For simplicity of specification, energy efficiency bands have been introduced based on the SEDBUK ratings. Most manufacturers use these ratings in their product literature, but it is not a requirement. The range bands (in percentage terms) are:

Excellent	Poor

A, 90 plus   B, 86–90   C, 82–86   D, 78–82   E, 74–78   F, 70–74   G, below 70

The database also contains values for many obsolete boilers, thus enabling cost comparisons to be made when considering whether to replace an old boiler. Typical factored differences based on a value of 1.0 for an efficient new condensing boiler are shown in Table 5.

Table 5
What is an efficient boiler? Typical comparison factors with a new condensing boiler (value 1.0)

Boiler type	Seasonal efficiency	Flat	Bungalow	Terraced	Semi-detached	Detached
Old heavyweight	55%	1.50	1.52	1.53	1.53	1.55
Old lightweight	65%	1.30	1.31	1.31	1.31	1.52
New non-condensing	78%	1.11	1.11	1.11	1.11	1.12
New condensing	88%	1.00	1.00	1.00	1.00	1.00

It therefore costs up to 55% more to operate an old heavyweight boiler in a typical detached house than a new condensing boiler in grade band B. In cash terms, this equates to an annual additional fuel cost of £90 for a flat and £195 for a detached house at mid-2005 prices.

SEDBUK ratings have been calculated for other types of fuel source and these are published in Table 4 of the Government's *Standard Assessment Procedure for Energy Rating of Dwellings* (SAP rating guide), which can be found at www.bre.co.uk/sap2005.

The preferred order for establishing the seasonal efficiency rating for a boiler is:[60,61]

1. for new and old gas- and oil-fired boilers, consult the SEDBUK database;
2. for new gas- and oil-fired boilers not listed in the SEDBUK database, obtain an independent certifier's certificate from the manufacturer if available;
3. if information on gas and oil boilers is not available using methods 1 and 2, and for all other fuel sources, use Table 4 in the SAP rating guide;
4. Add efficiency adjustments from Table 4c.

Reference will need to be made to the following documents for performance requirements as appropriate:

- *The Domestic Heating Compliance Guide* is a second-tier document which sets out clear guidance on means of complying with the requirements of Parts L1A and L1B with sections on gas- and oil-fired heating and hot water systems as well as electric, solid fuel, community, underfloor and heat pump heating systems. There is a section on solar heating and micro-combined heat and power. Once the method of providing hot water and heating has been decided, then the tables in each section provide the target performance and criteria for the boiler, cylinder, pipework, radiators and controls.
- *The Non-Domestic Heating, Cooling and Ventilation Compliance Guide* is a second-tier document that sets out clear guidance on means of complying with the requirements of Parts L2A and L2B, with sections on boilers, heat pumps, gas- and oil-fired warm air heaters, gas- and oil-fired radiant technology, combined heat and power (CHP), electric space heating, domestic (meaning domestic in type or size) hot water, comfort cooling, air distribution systems, pipework and duct insulation.

If equipment is being replaced, in non-domestic buildings there is scope to 'earn' heating efficiency credits to improve its effective heat-generating seasonal efficiency rating for incorporation into SBEM or another accredited modelling tool. A list of credit scores is included in Table 7 of *The Non-Domestic Heating, Cooling and Ventilation Compliance Guide* and they are awarded for reducing overdesign of boilers, introduction of weather compensation valves, optimised start/stop control systems, full building management systems and other such options.

Notes

60. BRE on behalf of DEFRA, *The Government's Standard Assessment Procedure for Energy Rating of Dwellings*, draft 2005 edition.

61. DEFRA, *Boiler Efficiency Database*. www.sedbuk.com

There is a compliance checklist in Appendix 2 of *The Non-Domestic Heating, Cooling and Ventilation Compliance Guide*.

6.10 What are efficient heating controls?

6.10.1 Non-dwellings

The Non-Domestic Heating, Cooling and Ventilation Compliance Guide refers to the 'minimum controls package', which represents the minimum requirement for compliance with Parts L2A and L2B.

The requirements for the minimum controls package are different for each type of system.

For example, Table 4 in *The Non-Domestic Heating, Cooling and Ventilation Compliance Guide* describes three packages, each applying to a different boiler plant output. As the output of the boiler increases, the controls requirement increases. A boiler of less than 100KW needs timing and temperature controls (in zones where the building area is greater than 150m^2) and weather compensation (except where constant temperature supply is required). But a larger boiler of less than 500kW also needs optimal start/stop controls with night set-back (or frost protection outside occupied periods) and two-stage high/low firing facility (or multiple boilers with sequence controllers).

6.10.2 Dwellings

Whatever the efficiency of the primary heating source, its efficiency can be further enhanced (or degraded) by the level of control available. The SAP energy rating guide identifies 15 types of control or system which, if used either singly or in combination, can improve performance. These are:

Room thermostat
Switches space heating on/off at a single adjustable temperature which is preset by the user.

Time switch
Controls either space heating or hot water for one or more on/off cycles per day.

Programmer
Time switch capable of controlling space heating and hot water independently.

Programmable room thermostat
Combined room thermostat and time switch.

Delayed start thermostat
Delays timed heating dependent on internal or external temperature.

Thermostatic radiator valve (TRV)
Controls heat output according to air temperature.

Cylinder thermostat
Switches water heating on/off at a single user adjustable temperature.

Flow switch
Switches system off when all TRVs are closed.

Boiler interlock
Control arrangement preventing boiler firing when there is no heat demand.

By-pass
Arrangement of pipes to ensure that minimum flow rate is maintained in the boiler.

Boiler energy manager
Vary in complexity from a simple device to delay boiler firing to full optimisation control.

Time and temperature zone controls
Minimum two with independent time and temperature control.

Weather compensator
Controls system water temperature according to external temperature.

Load compensator
Controls system water temperature according to internal temperature.

Controls for electric storage heaters:

- manual charge control – adjusted by user;
- automatic charge control – using internal or external temperature sensors;
- cellnet type controller – electronic control of heaters.

The beneficial effects of increased levels of control are accounted for in three ways:

1. An improvement to the SEDBUK rating. These range from +3% for a condensing boiler to –10% if there is no thermostatic control of room temperature (see Appendix 6.9).
2. The ability to maintain a greater temperature difference between living areas and the rest of the dwelling as derived from Table 9 of the SAP guide.
3. The ability to reduce the living area mid-internal temperature obtained from Table 8 of the SAP guide (Tables). (Lack of effective controls has the negative effect of requiring an increase in this temperature.)

The various combinations of control produce the following results in order of improvement for differing systems which are defined by the following groups:

1. boiler systems with radiators or underfloor heating;
2. heat pumps;
3. community heating schemes (seek specialist advice);
4. electric storage systems;
5. warm air systems;
6. room heater systems;
7. other systems.

All dwellings (except one-level open plan dwellings with more than 70% living area) should be divided into two heating zones with separate thermostats; each zone in dwellings greater than 150m² should also have a separate timing control.

Full requirements for various systems are contained in the second-tier publication *Domestic Heating Compliance Guide* (2006).[62]

Notes

62. *Domestic Heating Compliance Guide*, NBS, 2006.

Table 6
Table showing HLP values to be used in Table 9 of the SAP guide (Tables)

	Systems group					
A higher control value = A more effective system	1	2	4	5	6	7
No time or thermostatic control of room temperature	1*	1^		1^	1^	1^
Programmer, no thermostat	1	1		1		1
Room thermostat only	1	1		1		1
Programmer + room thermostat	1	1		1		1
Programmer + at least two room thermostats	2	2				
Programmer + room thermostat + TRVs	2					
Programmer + TRVs + by-pass	2	2				
Programmer + TRVs + flow switch or boiler energy manager	2					
Temperature zone control						2
Time and temperature zone control	3	3		3		3
Appliance thermostats					3	
Programmer + room or appliance thermostats					3	
Manual^ or automatic charge control or Celect type controls		3				

For control values marked * and ^ the living area mid-internal temperature obtained from Table 8 of the SAP guide should be modified by +0.6K and +0.3K respectively. This has a negative effect on the overall rating.

6.11 How to insulate pipes, ducts, tanks and cylinders

6.11.1 General

Pipes, ducts and equipment require insulation to: protect against heat loss; keep temperatures constant; minimise the risk of damage by frost or freezing, condensation or fire; and provide an element of acoustic protection and safety in keeping human contact with very hot or cold surfaces to a minimum.[63]

Notes

63. BS 5422:2001 Method for Specifying Thermal Insulating Materials for Pipes, Tanks, Vessels, Ductwork and Equipment Operating Within the Temperature Range –40K to +700K. BS 5970:2001 Code of Practice for Thermal Insulation of Pipework and Equipment in the Temperature Range of –100K to +870K.

The type of insulation will depend on the location of the equipment, i.e. whether it is internal or external or above or below ground.

For small- to medium-bore pipes (domestic or small commercial installations), a one-piece pre-formed insulation is appropriate. Examples of this may be wool, felt, cork, flexible rubber or fibreglass. Larger bore pipes over 250mm diameter will require a quilted material secured with metal bands or adhesive tape with taped joints.

Rectangular ducts and tanks should have a rigid or semi-rigid slab insulation, cut to fit with overlaps at angles and bonded as recommended by the manufacturer. Duct insulation should be covered with an impermeable finish to provide protection and shed condensation. In external situations, this should be a sheet metal protection for additional strength.

Cylinders can be insulated using the same method as tanks, although both tanks and cylinders in domestic situations can be fitted with a quilted jacket. New domestic cylinders are usually insulated with foam before they are installed.

6.11.2 Dwellings

Tables 3, 11, 20 and 25 of the *Domestic Heating Compliance Guide* describe the minimum provision for insulation of pipes serving new and replacement gas-, oil-, solid fuel-fired and community central heating systems. Primary circulation pipes should be insulated whenever they pass outside the heated living space, domestic hot water pipes should be insulated throughout their length (except where impractical – say through joists) and all pipes connected to hot water storage vessels should be insulated for 1m from their point of connection to the cylinder (or to where they become concealed). Lesser standards are acceptable for replacement systems where access is unavailable. Buried community heating pipes tend to be pre-insulated to EN 253, but variable volume controls assist by maintaining low return temperatures.

Table 33 of the *Domestic Heating Compliance Guide* requires that all primary circuit pipes should be insulated for solar hot water systems throughout their length. Pipes connected to hot water storage vessels are as described in the preceding paragraph.

All insulation should be labelled as complying with the *Domestic Heating Compliance Guide* (or lesser standard where there are constraints). The heat losses for various diameter pipes are included in Table 3 referred to above.

6.11.3 Non-dwellings

Pipes and ducts are insulated to avoid heat loss in heating systems and heat gain in cooling systems.

The *Non-Domestic Heating, Cooling and Ventilation Compliance Guide* requires all hot water and heating pipework to be insulated so that heat loss only occurs where 'it is useful'. Maximum heat loss requirements to determine required insulation levels are given in Table 37 of the compliance guide.

The cooling load to compensate for heat gain in distribution pipework should be less than 5% of the total load. Maximum heat gain requirements to determine required insulation levels are given in Table 38 of the compliance guide.

Where the same ducting is used for heating and cooling, it should meet the requirements for chilled ductwork in accordance with Table 39 in the compliance guide – heat loss/gain per unit area, heat gains being expressed as a negative value.

6.12 How to provide an acceptable ventilation system

6.12.1 General

The most energy-efficient solution is properly designed natural ventilation in a building with low air permeability. The *Guide to Part F* (NBS, 2006) describes minimum ventilation requirements. However, external noise or pollutants may mean a mechanical system will be needed with a consequent increase in energy consumption. Minimising internal and external heat gains is a key part of an energy-efficient ventilation strategy. (Refer to Appendix 6.7 for further information.)

6.12.2 Dwellings

The guidance for ventilation in dwellings in Parts L1A (paragraph 40) and L1B (paragraph 34) is published (although it may not be up to date) in GBG268 *Efficient Ventilation in Housing*.

Air leakage rates in whole-house ventilation systems need to be to a much higher standard than for local extract with background ventilation, with four air changes per hour recommended for the former at 50Pa and with 5–7 air changes per hour at 50Pa for the latter. (The units require conversion from air changes per hour to $m^3/h \cdot m^2$ if they are to be compared with the SAP default for TER of $10m^3/h \cdot m^2$.) There are five kinds of ventilation system referred to in the guide:

- passive stack ventilation (PSV);
- local extract fans;
- heat recovery room ventilators;
- mechanical supply ventilation;
- whole-house mechanical ventilation with heat recovery (MVHR).

Although whole-house mechanical ventilation systems with heat recovery have a high capital cost and need maintenance, they can reduce energy consumption in the winter months and assist with control of condensation.

6.12.3 Non-dwellings

The only specific information stated in both Part L2A and Part L2B is that the mechanical ventilation system should be capable of achieving a specific fan power at 25% of design flow rate no greater than that achieved at 100% design flow rate. Parts L2A and L2B state that reasonable provision for ventilation system fans rated at more than 1100W would be to equip them with variable speed drives.

More detail can be found in Section 10, Air Distribution Systems, in *The Non-Domestic Heating, Cooling and Ventilation Compliance Guide*, which describes the requirements for both central and local ventilation systems (with and without heating, cooling and heat recovery).

Table 36 of the compliance guide shows limiting-specific fan powers in W/litre second for various system types and relates to work in existing buildings. The values for new buildings are shown in Table 35 of the compliance guide. As a general rule, specific fan power (a measure of efficiency) should be as low as possible (below 2W/litre second is good practice in offices[64]). Fan power is a factor of energy modelling calculations; in SBEM there is an option to use the default, which is set at 3W/litre second.

Ductwork should have minimal leakage. A way of achieving this is to fabricate the ducts to the specifications given in HVCA DW144 *Specifications for Sheet Metal Ductwork* (1998). SBEM gives better results for calculations if leak tests for ductwork and air-handling units meet the CEN classifications.

There is specific guidance for hospitals and schools published by NHS and DfES respectively.

Part L requires that ventilation systems have effective controls, appropriate commissioning and the provision of operating systems and instructions. Information on appropriate controls for different sizes of plant are tabulated in *The Non-Domestic Heating, Cooling and Ventilation Compliance Guide*, NBS (2006).

6.13 What is an efficient air-conditioning system?

6.13.1 Dwellings

Although natural ventilation is the most energy-efficient way of conditioning internal air, it may be necessary to introduce mechanical air-

---

Notes

64. Jones, P. *Energy Efficient Ventilation and Air Conditioning*, Inside Energy: The Continuing Professional Development Programme, module 6. September 2003.

conditioning when the external environment (noise or pollution) makes it impractical to open windows.

All fixed air-conditioning units for use in dwellings are now required to be labelled with an energy rating. The label is similar to that seen on a new refrigerator or washing machine marked EU and graded from A to G. A represents the most energy efficient and G the least.

Parts L1A and L1B require that fixed air conditioners should have a rating equal to or better than class C.

Pipes as well as ducts carrying treated air should be insulated (Part L1A, paragraph 39, and L1B, paragraph 33).

It will also be necessary to consider the impact of internal and external thermal heat gains (see Appendix 6.7) and to meet the requirements for ventilation under Part F.

6.13.2 Non-dwellings

The emphasis is on reducing internal and external heat gains. Reference is made to BR 364 for guidance on solar control strategies (see also Appendix 6.7), reducing heat gains to cooling equipment, reducing air leakage from buildings, reducing air leakage from ductwork and providing suitable controls, appropriate commissioning and operating systems/instructions to enable energy-efficient operation of the building.

Parts L2A and L2B measure only carbon dioxide emission improvements against a notional building of the same size and shape as the proposed building, so there is no provision under the Building Regulations to consider amending the shape of the building as part of the strategy to reduce the energy consumption of air-conditioning systems.

The scope of Section 8 of *The Non-Domestic Heating, Cooling and Ventilation Compliance Guide* aligns guidance on all main types of cooling equipment with the descriptions used in the SBEM tool.

The efficiency of refrigeration plant is defined as the seasonal energy efficiency ratio (SEER), the ratio of total annual cooling energy provided divided by total annual energy input. Heat rejection equipment associated with cooling, evaporative cooling and desiccant cooling systems is not within the scope of the guidance.

There are various ways to calculate SEER for single and multiple chiller systems and when part load energy efficiency ratios are known (for offices only). These methods are described in the compliance guide and the result should be no worse than that shown in Table 31 of the compliance guide for the minimum SEER for comfort cooling.

6.13.3 System design

Natural ventilation is most energy efficient (note the improvement factors in Table 1a in Part L2A), but it may not be sufficient to meet comfort criteria where there are large glass façades, deep office spaces with high internal heat gains and so on (see paragraph 65 in Part L2A).

Each zone (defined by activity, internal or external thermal gains or independent conditioning system) should be considered independently.

> The designer may have to consider long-term use of the building and its ability to be readily adapted to meet the changing needs of the user. This will determine the requirement for in-built flexibility of air-conditioning systems.

The SBEM calculator includes various choices for air-conditioning systems and each of them has default values for the CO_2 emissions for a variety of fuel sources. The emission factors for a variety of fuel sources are given in Table 2 of Part L2A.

Improvements can be made in energy performance by using one or more of the following approaches:

- consider the use of heat recovery systems;
- the system itself should not be overdesigned and should be optimised for each zone;
- the design should take into account ease of access for repairs and replacements of plant and the controls;
- efficiency is optimised by use of variable speed driven primary and secondary pumps;
- paragraph 43(e) of Part L2B requires the provision of energy metering on newly installed plant;
- primary pipes containing conditioned fluid should be as short as possible and well insulated to avoid heat gain of coolant;
- ducts containing conditioned air need to be insulated to avoid heat gain; all ducts need to be designed to meet requirements for airtightness (see paragraph 45 of Part L2A and paragraph 60 of Part L2B);
- consider increasing temperature range in water and reducing flow rates (this reduces the energy used by pumps);
- consider improving air flow in (less resistance) duct work.

Enhanced capital allowances (ECAs) enable businesses to claim 100% first-year capital allowances on their spending on qualifying plant and machinery. Businesses can write off the whole of the capital cost of their investment in certain energy-conserving technologies against their taxable profits for the period during which they make the investment. Details can be found on www.eca.gov.uk.

Controls

- The system controls should be readily accessible so that they can be easily over-ridden and should run the plant only when it is required.
- Consider the use of refrigerant loss monitors; note that an annual check will be required when there is more than 3kg of refrigerant in the system.
- Table 3 of Part L2A gives beneficial adjustment factors for carbon dioxide emissions if automatic monitoring systems with alarms are incorporated into the design.

Commissioning

- Ductwork on systems served by fans with a design flow rate greater than 1m³/s (and also ducts designed for BER to have leak rates lower than standard) should be tested for leakage (paragraph 80 of Part L2A) in accordance with procedures set out in HVA DW/143 and by a suitably competent person (e.g. a member of the HVCA Specialist Ductwork Group or a member of the Association of Ductwork Contractors and Allied Services).
- A log book should be provided in accordance with paragraph 82 of Part L2A (a way of showing compliance is using CIBSE TM 31 *Building Logbook Toolkit*). The log book must include the information used to calculate TER and BER.
- For existing buildings under Part L2B, a new or updated log book should be provided that should include details of newly provided services, their operation and maintenance and any newly installed energy meters (see Appendix 6.18).

Operation and maintenance
Instructions should include:

- regular calibration of controls;
- regular cleaning of dirty filters – and clearing dust from fins;
- regular inspection of ductwork to ensure insulation and airtightness are maintained.

These checks aim to avoid too much deterioration in the energy performance of the system.

6.14 What are energy-efficient light fittings?

Replacing older switch-start fluorescent lighting with modern fittings that include high-frequency electronic ballasts, high-efficiency tubes and high-efficacy reflectors can typically reduce energy consumption by 35%.

Lighting ballasts is the generic term for electrical or electronic components that are required to control the current passing through fluorescent discharge tubes. These ballasts dissipate energy themselves and can affect the light output efficiency of the tube itself.

These measures are particularly relevant to non-domestic installations, although low-output fluorescent lamps can be used successfully in domestic situations, with the lamps lasting up to 12 times longer than the equivalent standard incandescent lamp.

Low-energy lamps such as compact fluorescents generally have an average light output equivalent to five times their power consumption, e.g. a low-energy lamp consuming 15W of power will have a light output in the region of 75W.

All four sections of Part L require fittings to be installed that can accept only low-energy lamps. Details of the requirements have been included in Chapters 2, 3, 4 and 5 of this guide.

Parts L2A and L2B state that it is reasonable provision to locate lighting controls where they would encourage occupiers to switch off lighting when there is sufficient daylight or the space is not in use. Local controls should be no more than 6m away from the luminaire (or a distance of twice the luminaire height from the floor if this is greater). Dimming should be by reduction (and not diversion) of the power supply.

6.15 How to calculate luminaire-lumens/circuit-watt

This calculation will not normally be required in order to comply with Parts L1A and L1B. For most domestic situations, it will generally be adequate to provide fittings that accept only energy-efficient light bulbs (see 2.2.6 and 3.2.8 of this guide). This will have the effect of limiting the luminaire efficacy (nlum) to 40 luminaire-lumens/circuit-watt. For Parts L2A and L2B, the calculation below may be used to calculate the value of nlum.

Two units have been used – 'luminaire-lumens/circuit-watt' and 'lamp-lumens/circuit-watt'. A luminaire contains one or more lamps housed in a fitting and care must be taken to ensure that the correct units are being applied.

A step-by-step guide follows.

For each luminaire:

1. establish the 'average initial (100 hour) lumen output' (phi lamp) for each lamp in the luminaire and add them together (A);
2. multiply (A) by the 'light output ratio' (LOR)* of the luminaire to give (B);
3. divide (B) by the 'luminaire control factor' (C_L) taken from the table to give (C).

For the building:

4. sum the values of (C) for each luminaire and divide by 'sum of the circuit-watts' (P) for all the luminaires to give the 'luminaire efficacy' (nlum).

*LOR is defined as the ratio of the total light output of a luminaire under stated practical conditions to that of the lamp or lamps under reference conditions.

Table 7
Luminaire control factors used to calculate luminaire-lumens/circuit-watt

Control function	C_L
Installations designed to meet Part 2LA	1.00
Installations designed to meet Part 2LB:	
a Luminaire in daylit space with photoelectric switching or dimming or local manual switching	0.80
b Luminaire in generally unoccupied space with manual on switching but sensor-controlled off switching	0.80
c Both (a) and (b) above are provided	0.75
d In all other circumstances	1.00

C_L, value of luminaire control factor.

6.16 What a Part L site checklist should look like

6.16.1 Designers' site stage checklist

The following checklist, which covers all types of new building, is intended to help designers but may not cover every situation that could arise. It is important to foresee the questions that might have to be answered in the checklists at the end of Parts L1A and L2A and plan for them from the commencement of the job. These are the kinds of questions that may be appropriate:

- Is the full design information that was used for the design stage energy assessments available on site?
- Has the 'SAP' or 'accredited energy' assessor for the project checked any changes to details and/or specification (thermal envelope and services)?
- Have arrangements been made for the 'SAP' or 'accredited energy' assessor to visit the site and sign off the construction as being compliant with the design as work progresses? (Before work critical to the energy performance of the building is covered by subsequent construction.)
- Who will keep a record of site checks and inspections? Are the records of checks made secure from loss?
- Are properly documented records of specific site checks and inspections being made on a regular basis as the works proceed? Records are likely to include the item being checked, the person undertaking the check, the result of the check and any remedial work undertaken. Items for specific attention on site include:
  - Location, thickness and method of fixing in position for thermal insulation.
  - Gaps at edges and junctions in thermal insulation.
  - Thermal bridging (not anticipated in the design).
  - Location, continuity and fixing of vapour barriers and membranes.
  - Airtightness of components, windows, doors, walls, ceilings and floors.
  - Airtightness at junctions between components:
    wall to window/door;
    wall to roof;
    wall to floor;
    junctions between claddings.
  - Effective sealing around pipes and other service entries. This is particularly critical where the work will be concealed behind dry-lining, ductwork, kitchen units or other fitted furniture.
  - Are enclosing masonry walls sealed by wet plastering or by a high quality of jointing and pointing? (Some types of building block are inherently porous and must be plastered or sealed.)
  - Are the service installations, heating, ventilation, lighting (and any air-conditioning) as intended in the design?
  - Are passive solar control measures (if any) being constructed as intended in the design?
  - Are any walls separating dwellings from common parts of the building constructed as designed?
- Does air-handling ductwork need to be pressure tested and, if so, who will undertake the test? (This only applies to non-domestic installations.)
- Who will be the 'approved competent' person to sign off the lighting installation as being compliant with the design?
- Will on-site air pressure testing of the building envelope be required, and who will undertake the tests?
- Who will carry out commissioning of building services and certify performance?
- Who is responsible for collecting and collating the information on operation and maintenance to be passed to the building's user/owner?

The standard of site checking required is much more onerous than that traditionally required of building designers. For new buildings, reference should be made to the checklist included as Appendix A to the appropriate part of Part L. Although other methods of recording could be used, any alternative should be agreed with the Building Control Body before construction.

6.16.2 What site checks should be made and by whom?

L1A and L2A include a detailed list (Appendix A) of specified checks to be made and signed off to demonstrate compliance. These are tabulated under the following headings:

Check   Evidence   Produced by   Design OK?   As-built OK?

Lead designers may wish to confirm, in specifications, which checks are to be made and by whom. The Appendix A checklists will draw together checks made by the following persons:

- SAP assessor (accredited FAERO/BRE Certification model assessor for non-domestic);
- builder or electrical contractor;
- developer (buildings that are not dwellings);
- BINDT registered member (air permeability test);
- approved competent person (heating and hot water, electric and lighting systems);
- suitably qualified person (services commissioning);
- suitably qualified person (ductwork air test).

It may be sensible also to include the Building Control body in the pre-planned checking process.

On large developments, keeping track of records will be a significant task and will need to be co-ordinated at design stage and on site. Numerous items require the energy assessor to confirm that the design is satisfactory and that the 'as-built' construction is in accordance with the design.

6.17 How is a system properly commissioned?

6.17.1 Dwellings

For the construction of a new dwelling, extension of a dwelling, installation of a new system and modification to existing systems, the provisions are identical.

The heating and hot water system(s) should be commissioned so that at completion the system(s) and their controls are left in working order and can operate efficiently for the purpose of the conservation of fuel and power (see also Appendix 6.2.8 of this guide).

Advice is available in the *Domestic Heating Compliance Guide*, NBS (2006).

> An approved competent person must certify commissioning of heating and hot water systems. The person carrying out the work must provide the local authority with a notice confirming that all fixed services have been properly commissioned in accordance with a Government-approved procedure.

6.17.2 Non-dwellings

New buildings and works to existing buildings will require commissioning of all **controlled services** (heating, hot water, electrical and mechanical) in accordance with CIBSE Code M by a suitably qualified person, e.g. a member of the Commissioning Specialists Association or the Commissioning Group of the HVCA.

All ductwork must be tested for air leakage by a suitably qualified person, e.g. a member of the HVCA Specialist Ductwork Group or the Association of Ductwork Contractors and Allied Services.

> An approved competent person must certify commissioning of *controlled* **services**. The person carrying out the work must provide to the local authority a notice confirming that all fixed services have been properly commissioned in accordance with a procedure approved by the Government.

The requirement for non-dwellings is more onerous than for dwellings as it includes all services and not just heating and hot water.

6.18 How to compile an instruction/completion pack

6.18.1 New dwellings

Information must be provided to *each dwelling owner*, as required by Part L1A paragraphs 67 and 68. However, the building 'occupier' is to be provided with 'sufficient information, including operating and maintenance instructions, enabling the building and the building services to be operated and maintained in such a manner as to consume no more fuel and power than is reasonable in the circumstances'. It would therefore be appropriate that both owner and occupier (if they are not the same person or organisation) are provided with suitable operating instructions as part of the home improvement pack.

The actual information to be supplied will depend on the complexity of the dwelling, but the following list should cover most requirements.

- Heating system
- Adjustment of boiler temperature (or other heat source)
- Setting operating times
    - Adjustment of room temperatures (temperature-regulating valves and room thermostat)
    - Boiler service interval
    - Boiler manufacturer's user instructions
- Hot water
    - Adjustment of water temperature
    - Setting operating times
    - Avoidance of waste
- Ventilation
    - Use of any mechanical and passive ventilation systems
    - Service intervals for mechanical ventilation
    - Cleaning or replacement of filters
- Lighting
    - Use of energy-efficient lighting
    - Life of lamps and replacement procedure
- Solar gain
    - Use and maintenance of any solar protection measures provided
- SAP
    - Predicted SAP score, energy usage and annual CO_2 output

Note that the energy rating of the dwelling, its SAP score, should be prepared and displayed in a conspicuous place (paragraph 70 of Part L1A).

6.18.2 Works to existing dwellings

Similar requirements apply as for new dwellings in relation to the scope of works as undertaken. A SAP assessment is not required by Part L1B, except for a **material change of use**, but may be needed in future as part of sale documentation or the home improvement pack.

6.18.3 Non-dwellings

An operating and maintenance manual (log book) is required. CIBSE publication TM31 provides guidance and standard templates for the content.

Designers should ensure that provision of the operating and maintenance manual is included in the contract specification or the M&E design brief.

6.19 How to assess 'simple payback' over time

When extending a building, or renewing or refurbishing a thermal element (wall, roof or floor), there may be situations where full compliance with an upgrade of U-value is not economic (paragraphs 54–57 and 72 of Part L1B and paragraphs 87–90 and 109 of Part L2B).

If the simple payback period is more than 15 years, the upgrade is not required. If the remaining life of the building is less than 15 years, payback is to be calculated only for the remaining life of the building.

To calculate the payback period:

$$\text{The payback period in years} = \frac{\text{Cost of additional insulation}}{\text{Estimated annual energy saving}}$$

Table A1 of Part L1B gives advice on what is likely to be economic where there is little or no existing thermal insulation material provided to the existing thermal element.

Where some thermal insulation is already provided or the upgrade will be unusually expensive, the payback period can be calculated to check if the upgrade is required.

Paragraph 72 of Part L1B and paragraph 109 of Part L2B state that prices should be current at the date that the proposals are made known to the Building Control body, and confirmed in a report signed by a suitably qualified person (say a chartered quantity surveyor). Energy prices in pence per kilowatt-hour are included in the relevant clauses for the purpose of the calculation, although there is an option for dwelling owners or building owners to adjust the cost of energy upwards. The cost including labour and materials can be calculated from contract rates or published price guides.

The estimated annual saving can only be obtained by use of SAP for dwellings or an SBEM or an approved modelling tool for non-dwellings. This may be an additional project cost greater than any potential saving by not insulating.

The requirement to upgrade applies to any refurbishment which changes the thermal performance of more than 25% of a wall, roof or floor. The requirement does not apply to redecoration.

Reference documents and what they cover

7

TABLE 8
Documents referenced in Part L

Document	Synopsis	Reference
AM10 *Natural Ventilation in Non-Domestic Buildings*, CIBSE (2005)	Guidance on natural ventilation, but still refers to the previous version of Part L	L2A
BRE Digest DG 498 *Selecting lighting controls* (March 2006)	This Digest explains the common types of control, when to use them and how to calculate energy savings	L2A and L2B
BRE Good Building Guide 37 *Insulating roofs at rafter level: Sarking insulation*	Guidance on good practice in placing insulation to roofs at high level	L1B
BRE Information Paper IP 1/06 *Assessing the effects of thermal bridging at junctions and around openings* (2006)	Guidance on thermal bridging and air leakage	L1A and L2A
BRE Report BR 262 *Thermal insulation: avoiding risks*, BRE (2001)	Provides detailed guidance by building element on avoiding the risks associated with higher levels of insulation. It includes suggested details for a number of locations	All Parts
BRE Report BR 364 *Solar shading of buildings 2001* (1999)	Guidance on effectiveness of different solar shading solutions	L2A and L2B
BRE Report BR 443 *Conventions for U-value calculations*, BRE (2006)	Guidance on U-value calculations	All Parts
BRE Report BR 448 *Airtightness in commercial and public buildings* (2002)	Outline guide to design, setting out the principles of providing an effective airtightness layer and advises on some common pitfalls	L2A
BS 5250:2002 Code of practice for control of condensation in buildings.	Guidance on the principles and practice of locating insulation and arranging ventilation to reduce the risk of condensation	L1B
BS 5803-5:1985 Thermal insulation for use in pitched. roof spaces in dwellings.	Code of practice including suggested details for mineral and cellulose fibre insulation within the roof space	L1B
BS 6229:2003 Flat roofs with continuously supported coverings. Code of practice.	General code of practice, but makes reference to thermal design and avoidance of condensation	L1B

Table 8
Continued

BS 8206-2:1992 Code of practice for daylighting.	Guidance on providing adequate daylight while reducing gains by reducing window sizes	L1A and L2A
BS EN ISO 13788:2002 Hygrothermal performance of building components and building elements.	Provides calculation methods to identify risks of condensation, both interstitial and surface	L1B
Building Bulletin BB87 *Guidelines for Environmental Design in Schools*, DfES (2003)	Schools design guidance, but currently deals with the previous version of Part L	L2A
Building Regulations and historic buildings. Interim Guidance Note, English Heritage (2002)	Guidance on the application of Part L to buildings which are listed or have architectural or historic interest	L1B and L2B
CE129 *Reducing overheating – a designer's guide*, Energy Saving Trust (2005 edition)	Guidance on how to reduce the risk of overheating in summer in dwellings	L1A
CE66 *Windows for new and existing housing*, Energy Saving Trust	Best practice guidance for window specification	L1B and L2B
Code of practice for loft insulation, National Insulation Association.	Guidance on recommended practice for loft insulation	L1B
Commissioning Code M: Commissioning Management, CIBSE (2003)	Commissioning procedure for non-domestic building works	L2A and L2B
CWCT/CAB report on curtain walling	To be issued	L2A and L2B
Domestic Heating Compliance Guide, ODPM/NBS (2006)	Describes requirements for compliance with L1A and L1B for heating and hot water services and associated controls and pipework	L1A and L1B
Draft Building Bulletin 101 *Ventilation of School Buildings*, DfES	BB101 gives guidance on providing ventilation in school buildings in line with the 2005 revision of Part F	L2A
DW/143 *A Practical Guide to Ductwork Leakage Testing*, HVCA (2000)	Procedure required with fan flow rates greater than $1m^3/s$	L2A and L2B
Energy performance standards for modular and portable buildings, MPBA (2006)	A publication of the Modular and Portable Building Association in relation to Part L	L2A
Energy Saving Trust guides: CE17, 57, 58, 59, 83, 97 and GPG 294, 296, 297	General guidance on renovation options available for walls, roofs and floors	L1B

GIL 20 *Low energy domestic lighting*, EST (2006)	A guide to the type of low-energy light fittings which are available where they could be used and how to specify them	L1A and L1B
GIL 20 *Low energy domestic lighting. The benefits of compact fluorescent lamps in housing – a monitored study*	Guide to low-energy light sources for domestic use	L1A and L2A
GIL 65 *Metering energy use in non-domestic buildings* (2004)	Recommendations on metering to allow identification of energy use in different systems	L2B
GPG 268 *Energy efficient ventilation in dwellings – a guide for specifiers on the requirements and options for ventilation*, EST (2006)	Explains why ventilation is important, the impact that good ventilation has on achieving the efficient use of energy, and the importance of airtightness. It describes the advantages and disadvantages of a range of ventilation systems	L1A and L2A
Guide A: *Environmental design*, CIBSE (2006) www.cibse.org	The basic services design handbook, covering a wide range of topics	L2A
HVAC *Guidance for achieving compliance with Part L of the Building Regulations*, TIMSA (2006)	A reference for all compliance issues related to the treatment of pipework and ductwork	All Parts
HVCA DW/144 *Specifications for sheet metal ductwork* (1998)	Provides a means to demonstrating compliance with air leakage requirements	L2A and L2B
Limiting Thermal Bridging and Air Leakage: Robust Construction Details for Dwellings and Similar Buildings, Amendment 1, TSO (2002)	Guidance on thermal bridging in dwellings	L1A and L1B
Low or Zero Carbon Energy Sources: strategic guide, ODPM (2006)	Report on the suitability of alternative energy sources to reduce reliance on fossil fuels	L1A, L2A and L2B
Measuring Air Permeability of Building Envelopes, ATTMA (March 2006)	Standards and methods for testing air permeability of buildings	L1A and L2A
National Calculation Methodology, ODPM/NBS (2006)	Due to be published September 2006	L2A
Non-domestic heating, cooling and ventilation compliance guide, ODPM/NBS (2006)	Describes requirements for compliance with L2A and L2B for heating, hot water, cooling and ventilation services, and associated controls and pipework	L2A and L2B

SBEM user manual and calculation tool, BRE (2006). www.ncm.bre.co.uk	Calculation tool to demonstrate the energy requirements of a building using the Simplified Building Energy Method	L1A and L2A
Statutory Instrument SI 2005/1726 The Energy Information (Household Air Conditioners) (No. 2) Regulations 2005. www.opsi.gov.uk	Requirement for all household air-conditioning units to be labelled showing their energy efficiency	L1A and L2A
The Government's Standard Assessment Procedure for energy rating of dwellings, BRE/Defra (2005). www.bre.co.uk/sap2005	Guidance on (SAP) calculation of CO_2 emissions in domestic buildings	L1A
Thermal Insulation of H & V Ductwork, TIMSA (1998)	Sets the standards for insulating pipes, ducts and vessels	L2A and L2B
TM 36 *Climate change and the internal environment: Impacts and adaptation*, CIBSE	Guidance on the need to improve the passive performance of buildings to reduce the need for cooling in future	L1A and L2A
TM31 *Building logbook toolkit*, CIBSE (2006)	Guide to a compliant building log book, showing data required	L2A and L2B
TM33 *Standard tests for the assessment of building services design software*, CIBSE	This comprises a series of tests to be used to check that the software for building services design provides reliable results. It is to be updated if it is to be used in place of the non-domestic calculation methodology	L2A
TM37 *Design for improved solar shading control*, CIBSE (2006)	Guidance on the design of façades to incorporate appropriate levels of solar shading, and information on some design options	L2A
TM39 *Building Energy Metering (a guide to energy sub-metering in non-domestic buildings)*, CIBSE	An updated version of General Information Leaflet 65	L2A and L2B
TP17 *Guidance on design of metal roofing and cladding to comply with Approved Document L2*, MCRMA (2006)	To be issued (2002 version can be downloaded from MCRMA website: www.mcrma.co.uk)	L2A and L2B
Use of rooflights to satisfy the 2002 Building Regulations for the conservation of fuel and power, NARM (2002)	Covers all aspects of rooflight design in relation to conservation of energy	All Parts

Glossary of terms

Accredited details	Published details that have been assessed and shown to meet the requirements of Part L in terms of air permeability, reduction of thermal bridging and condensation. Use of these details reduces the requirements for air permeability testing. At the time of going to press, no firm decision has been made on the adoption of accredited details. For a period following publication of Part L, buildings will be tested, and the quality of detailing assessed. This feedback will inform the design of the accredited details. In the meantime, details will have to be designed and constructed with a view to ensuring that air permeability, thermal bridging and condensation control measures meet the requirements of the Approved Document (refer also to IP1/06 and BR 262). Some manufacturers are beginning to publish details.
Accredited energy surveyor	An authorised SAP assessor (refer to this glossary).
Approved calculation tool	For dwellings of area less than 450m^2, the approved calculation tool is SAP. In other buildings and dwellings of area greater than 450m^2, this means SBEM or another calculation tool that is accredited as complying with the national calculation methodology. Refer to Schedule A reprinted in the front of each of the Part L Approved Documents.
Approved competent person	For ductwork leakage testing, a member of the HVCA specialist ductwork group or a member of the Association of Ductwork Contractors and Allied Services. For demonstration of simple payback or cost of consequential improvements, the equivalent of a chartered quantity surveyor Part J and Part P heating and hot water systems, authorised competent persons self-certification scheme (i.e. Corgi-registered plumber, ELECSA, the recognised competent person's scheme for AD P of the Building Regulations).
ATTMA	Air Tightness Testing and Measurement Association (a special interest group within the British Institute of Non-Destructive Testing.)
Authorised SAP assessor	A person who has met the requirements of FAERO (or their members Elmhurst Energy, MVM Consultants, National Energy Services) and is qualified to issue SAP ratings on NHER scheme certificates.

Building emission rate (BER)	The amount of carbon dioxide in kilograms emitted per square metre of floor area per year as the result of the provision of heating, cooling, hot water, ventilation and internal fixed lighting. This may be in the form of a prediction at design stage or a prediction based on the as-built dwelling, taking into account the results of air permeability tests. It is calculated using the SAP energy calculation tool (for details refer to Appendix 6.3).
Building work	In the context of the Building Regulations, 'building work' means: a. the erection or extension of a building; b. the provision or extension of a controlled service or fitting in or in connection with a building; c. the material alteration of a building, or a controlled service or fitting, as mentioned in paragraph 2; d. work required by regulation 6 (requirements relating to material change of use); e. the insertion of insulating material into the cavity wall of a building; f. work involving the underpinning of a building. See also material alteration.
Change in energy status	This seems to be a 'catch all' term to cover the change in thermal requirements that arises when there is no 'material change of use' within the meaning of the regulations.
Commissioning specialist	Member of the Commissioning Specialists Association or the Commissioning Group of the HVCA.
Consequential improvement works	Work that is triggered as a requirement to the existing building as a consequence of: • extending an existing building over 1000m²; or • providing fixed building services; or • increasing the installed capacity of an existing building service. The works which trigger the need for consequential works are called the principal works. The cost of consequential works should meet the requirements of simple payback calculations and should not be less than 10% of the cost of the principal works (unless the existing building already meets the required thermal performance).
Conservatory	An extension to a building that has not less than three-quarters of its roof area and not less than one-half of the external wall area made from translucent material, and is thermally separated from the building by walls, windows and doors with the same U-value and draught-stripping provisions as provided elsewhere in the building. (Please also refer to the definition of 'Extension to a dwelling' in this glossary.)

Controlled fittings

Controlled fittings are windows (including the glazed elements of a curtain wall), rooflights and doors (including high usage doors, large access doors for vehicles and roof ventilators) but not display windows.

Controlled services

Controlled services are heating and hot water systems, pipes and ducts, mechanical ventilation or cooling, fixed internal lighting including display lighting and occupier-controlled external lighting. It does not include emergency or specialist process lighting; refer to 'Specialist lighting' in this glossary for more detail.

Dwelling

A self-contained unit designed to accommodate a single household. Rooms for residential purposes are not dwellings so Part L2A is applicable to their construction.

Dwelling emission rate (DER)

The amount of carbon dioxide in kilograms emitted per square metre of floor area per year as the result of the provision of heating, hot water, ventilation and internal (and occupier-controlled external) fixed lighting. It can be calculated on a per dwelling basis or averaged over all the dwellings in a block. DER is both a prediction at design stage and a subsequent prediction based on the as-built dwelling taking into account the results of air permeability tests. It is calculated using the SAP energy calculation tool (for details, refer to Appendices 6.2 and 6.4).

Extension to a dwelling

An addition to a dwelling. There are exemptions that need not comply with the Building Regulations, and these are extensions of a building by the addition at ground level of:

a. a conservatory, porch, covered yard or covered way; or
b. a carport open on at least two sides;

where the floor area of that extension does not exceed 30m^2; in the case of a conservatory or a porch of less than 30m^2, the glazing should still satisfy the requirements of Part N of Schedule 1, Protection against impact, in particular, N1: 'Glazing, with which people are likely to come into contact whilst moving in or about the building shall (a) if broken on impact, break in a way which is unlikely to cause injury; or (b) resist impact without breaking; or (c) be shielded or protected from impact'.

Fit-out works

The fitting out of the shell of a building for occupation as an independently procured piece of work. The first fit-out of a building should be treated as part of the new works even if there is a delay between completing the shell and the commencement of the fit-out works.

Historic buildings

Special considerations (relaxations) apply to historic buildings, including listed buildings, buildings that are referred to in the local authority's development plan as being of local architectural or historic interest, or those buildings in national parks, areas of outstanding natural beauty and world heritage sites. When dealing with an historic building, refer to the English Heritage Guidance Note *Building Regulations and Historic Buildings*, English Heritage (September 2002).

Home Information Pack	From 1 June 2007, all home owners in England and Wales will need to prepare a Home Information Pack before putting a home up for sale. Before this date, the requirement to provide an energy certificate still applies. Refer to www.homeinformationpacks.gov.uk.
Low and zero carbon energy supply systems (LZC)	Such as absorption cooling, ground cooling, biomass, combined heat and power, ground source heat pumps, wind energy, photovoltaic and solar hot water.
Material alteration	A material alteration only arises if you make things worse in terms of compliance with certain parts of the Building Regulations. This could arise 'at any stage' in the works including (presumably) construction.

An alteration is material for the purposes of these Regulations if the work, or any part of it, would at any stage result in:

a. a building or controlled service or fitting not complying with a relevant requirement where previously it did; or
b. a building or controlled service or fitting that before the work commenced did not comply with a relevant requirement, being more unsatisfactory in relation to such a requirement.

In connection with the above 'relevant requirement' means any of the following applicable requirements of Schedule 1, namely:

Part A (structure)
    paragraph B1 (means of warning and escape)
    paragraph B3 (internal fire spread – structure)
    paragraph B4 (external fire spread)
    paragraph B5 (access and facilities for the fire service)
Part M (access and facilities for disabled people). |
| Material change of use | For the purposes of the Regulations there is a material change of use where there is a change in the purposes for which, or the circumstances in which, a building is used, so that after that change:

a. the building is used as a dwelling, where previously it was not (use Part L1B)
b. the building contains a flat where previously it did not (use Part L1B)
c. the building is used as an hotel or boarding house when previously it was not (use Part L2B)
d. the building is used as an institution where previously it was not (use Part L2B)
e. the building is used as a public building where previously it was not (use Part L2B)
f. the building is not a building described in Classes I to VI in Schedule 2, where previously it was (Part L no longer applies)
g. the building, which contains at least one dwelling, contains a greater or lesser number of dwellings than it did previously (use Part L1B) |

h. the building contains a room for residential purposes where previously it did not (use Part L1B if the room is a dwelling, e.g. a caretaker's flat; use Part L2B if the residential use is not a dwelling, such as a hostel)

i. the building, which contains at least one room for residential purposes, contains a greater or lesser number of rooms than it did previously (use Part L1B if the room is a dwelling, e.g. a caretaker's flat; use Part L2B if the residential use is not a dwelling, such as a hostel)

j. the building is used as a shop where previously it was not (use Part L2B).

Principal works	The main works. The works which trigger the need for consequential improvement works.
Qualified inspector	The National Association of Professional Inspectors and Testers provides an independent professional trade body for electrical inspectors, electrical contractors, electricians and allied trades throughout the UK. The National Inspection Council for Electrical Installation Contracting is the industry's independent, non-profit-making, voluntary regulatory body.
Reasonable provision	The ADs were originally conceived to set down reasonable means to provide compliant solutions under the Building Regulations, known in shorthand as 'reasonable provision'. These provisions are not mandatory if alternative solutions can be demonstrated to meet the requirements of the Building Regulations. Bear in mind that the carbon dioxide emission target to be achieved is mandatory.
Renovation	In relation to a roof, wall or floor, this means the provision of a new layer or replacement of an existing layer in the roof, wall or floor. It does not include decorative finishes.
Second-tier documents	Documents such as the *Domestic Heating Compliance Guide* that contain the detailed requirements referred to in the Approved Documents (the first-tier documents).
Simple payback	Means the amount of time it will take to recover the initial investment through energy savings. It is calculated by dividing the marginal additional cost of implementing an energy efficiency measure by the value of energy savings achieved by that measure. (This is based on current prices excluding VAT calculated by a suitably qualified person, e.g. a chartered quantity surveyor.) In the definitions given in Section 4 of Part L1B, there is an example of how to work out simple payback (see also Appendix 6.19).
Specialist lighting	This is excluded from the requirement to provide energy-efficient solutions. Specialist lighting includes theatre spotlights, projection equipment, TV and photographic studio lighting, medical lighting in operating theatres and doctors' and dentists' surgeries, illuminated signs, coloured or stroboscopic lighting, and art objects with integral lighting such as sculptures, decorative fountains and chandeliers.

Substantially glazed extension	An extension to a building that has less than three-quarters of its roof area and/or less than one-half of the external wall area made from translucent material, but unlike a normal extension is thermally separated from the main building by walls, windows and doors with the same U-value and draught-stripping provisions as provided elsewhere in the main building (please also refer to the definition of 'Extension to a dwelling' in this glossary).
Suitably qualified person	Suitably qualified persons to certify compliance for most aspects of a building are identified on the ODPM (DCLG) website. This does not include all aspects of a building which require certification by a 'suitably qualified person', for example, confirmation that there is no thermal bridging.

The Compliance Checklist for a new dwelling is included as Appendix A of L1A. Items 4.3 and 4.4 of that Appendix require the 'Builder' to provide evidence that 'Details conform to the standards set out in IP/106'. Clause 2 of Appendix A states that evidence should be provided by a 'suitably qualified person' and this 'may be accepted at face value at the discretion of the Building Control body dependent on the credentials of the person making the declaration'.

The Compliance Checklist for a new building that is not a dwelling is included as Appendix A of L2A. Items 4.2 and 4.3 of that Appendix require the 'Developer' to provide evidence that thermal bridging is acceptable with supporting documentary evidence'. Clause 2 of Appendix A states that evidence should be provided by a 'suitably qualified person acting for the developer' and this 'may be accepted at the discretion of the Building Control Body dependent on the credentials of the person making the declaration'.

The author recommends that to avoid any difficulties at practical completion, the person is identified and their status agreed with the Building Control body in advance of the construction phase. |
Target emission rate (TER)	The definition of this is different for dwellings and non-dwellings. It is the target amount of carbon dioxide in kilograms emitted per square metre of floor area per year as the result of the provision of heating, hot water, cooling ventilation and internal fixed lighting based on the carbon dioxide emissions from a notional building of the same size and shape as proposed with an improvement factor applied. In non-dwellings, the TER can also be adjusted to take account of low or zero energy carbon energy sources (for details refer to Appendix 6.1).
Thermal element	A thermal element is a wall (including the opaque elements of a curtain wall), floor, ceiling or roof that separates internal conditioned space from the external environment.
Third-tier documents	Documents such as industry guidance, good practice guides, codes of practice and standards.

Useful floor area   This equates to the gross floor area as measured in accordance with the
guidance issued by the RICS. It is described in the approved documents as
the total area of all enclosed spaces measured to the internal face of the
external walls. The area of sloping surfaces such as staircases, galleries,
raked auditoria and tiered terraces should be taken as their area on plan.
It should include areas occupied by partitions, column chimney breasts
and internal structural or party walls. It excludes areas not enclosed such
as open floors, covered ways and balconies (paragraph 112, L2B).